THE SOUL OF A
NEW MACHINE

ALSO BY TRACY KIDDER

The Road to Yuba City

THE SOUL
OF A NEW
MACHINE

Tracy Kidder

An Atlantic Monthly Press Book

Little, Brown and Company   Boston/Toronto

Third Printing

A portion of this book, in different form, first appeared in *The Atlantic.*

LIBRARY OF CONGRESS CATALOGING IN PUBLICATION DATA

Kidder, Tracy.
  The soul of a new machine.
  "An Atlantic Monthly Press book."
  1. Computer engineering—Popular works. 2. Data
General Corporation. I. Title.
TK7885.4.K53   621.3819′582   81-6044
ISBN 0-316-49170-5   AACR2

ATLANTIC–LITTLE, BROWN BOOKS
ARE PUBLISHED BY
LITTLE, BROWN AND COMPANY
IN ASSOCIATION WITH
THE ATLANTIC MONTHLY PRESS

BP
Designed by Susan Windheim
Published simultaneously in Canada
by Little, Brown & Company (Canada) Limited

PRINTED IN THE UNITED STATES OF AMERICA

To Richard Todd

THE SOUL OF A
NEW MACHINE

A GOOD MAN
IN A STORM

ALL THE WAY to the horizon in the last light, the sea was just degrees of gray, rolling and frothy on the surface. From the cockpit of a small white sloop — she was thirty-five feet long — the waves looked like hills coming up from behind, and most of the crew preferred not to glance at them. There were no other boats in sight, but off to the south for a while they could see the reassuring outlines of the coast. Then it got dark. Running under shortened sails in front of the northeaster, the boat rocked one way, gave a thump, and then it rolled the other. The pots and pans in the galley clanged. A six-pack of beer, which someone had forgotten to stow away, slid back and forth across the cabin floor, over and over again. Sometime late that night, one of the crew raised a voice against the wind and asked, "What are we trying to prove?"

All of them were adults. The owner and captain was a lawyer in his sixties. There were a psychologist and a physician and a professor, all of them in their late thirties, and also a man named Tom West. West was rather mysterious, being the merest acquaintance to one of them and a stranger to the others. They were bound for New York from Portland, Maine, on yachtsman's business, which is to say, primarily for sport. And when they had set sail in sheltered Casco Bay earlier that evening, decked out bravely in slick-

ers and sou'westers, all of them had felt at least a little bit romantic. But when they cleared the lee of the land and entered the sea-
way, and the boat suddenly began to lurch, they grabbed the
nearest sturdy things and thought about their suppers, which by
the time it got dark, several of them had lost.

Most of the crew now fell into that half-autistic state that the
monotony of storms at sea occasionally induces. You find a place
to sit and getting a good hold of it, you try not to move again. The
boat rolls this way and you flex the muscles around your stomach,
then relax; she rolls that way and you flex again. Just staying in
one place is exercise. For a while your mind may rebel: "Why did
you come, idiot? You don't have to be out here." You may feel
remorse for having cursed some part of life on land. After a time,
though, phrases start falling from your memory — snatches of
song or prayer or nursery rhymes — and you repeat them silently.
A little shot of spray in the face, however, or an especially loud
and dangerous-sounding thump from the hull, usually breaks the
trance and puts you back at sea again. You feel like a lonely child.
The ocean doesn't care about you. It makes your boat feel tiny.
The oceans are great promoters of religion, or at least of humil-
ity — but not in everyone.

In the glow of the running lights, most of the crew looked like
refugees, huddled, wearing blank faces. Among them, Tom West
appeared as a thin figure under a watch cap, in nearly constant
motion. High spirits had apparently possessed him from the moment they set sail, and the longer they were out in the storm, the
heavier the weather got, the livelier he grew. You could see him
grinning in the dark. West did all that the captain asked, so cheerfully, unquestioningly and fast, that one might have thought the
ghost of an old-fashioned virtuous seaman had joined them. Only
West never confessed to a queasy stomach. When one of the
others asked him if he felt seasick too, he replied, in a completely
serious voice, that he would not let himself. A little later, he made
his way down to the cabin, moving like a veteran conductor in a
rocking, rolling railroad car, and got himself a beer.

West was at the helm, the tiller in both hands, riding the waves; he was standing under a swaying lantern in the cabin studying the chart; he was nimbly climbing out onto the foredeck to wrestle in a jib and replace it with a smaller one. And when the captain decided to make for shelter, very late that night, at a little harbor with a passage into it that was twisty, narrow and full of tide, it was West, standing up in the bow, who spotted each unlighted channel marker and guided them safely in.

By dawn, the wind had moderated slightly and everyone felt better. They went out and raised their spinnaker. West gazed up at the large billowing sail and said, "The spinnaker looks like a win." He said, "Hey, we're haulin' ass." There was something faintly ridiculous about his exclamations, but also something childlike that made his companions smile. He was grinning most of the day, a cockeyed little smile that collected in one corner of his mouth. When the captain remarked worriedly that his boat had never gone so fast before, West laughed. He made the sound mostly in his throat. It was a low and even noise. Odd in itself and oddly provoked, the kind of laughter that ghost stories inspire, it seemed to say, "Here's something that's not ordinary."

A snapshot taken of the cockpit in the afternoon shows West sitting in the stern. The dark shadow of a day's growth of beard reveals that he passed adolescence some years ago, though just how many would be impossible to say. In fact, he is just forty. He wears glasses with flesh-colored rims, and a heavy gray sweater that must have given him long faithful service hangs loosely on his frame. He looks as if he must smell of wool. He looks thin, with a long narrow face that on a woman would be called horsey. A mane of brown hair, swept back behind his ears, reaches almost to his collar. His face is lifted, his lips pursed. He appears to be the person in command.

One of the crew would remember being alone with him on watch one night. They were sailing under clear skies with a gentle breeze. Suddenly, at the slackening of the tide, the wind fell away, some clouds rolled in, and then just as suddenly, when the tide began to run, the sky cleared up and the breeze returned. In a low and throaty voice, West made exclamations: "Did you see *that?*" He made his low and spooky laugh. His companion was about to say, "Well, I've seen this happen before." The tone of West's voice prevented him, however. He thought it would be rude to describe this event as ordinary. Besides, West was right, wasn't he? It *was* strange and wonderful the way the pieces of the weather sometimes played in concert. At any rate, it was fun to think that they had just encountered a natural mystery, and, somewhat surprised at himself, West's companion suggested that events like that made superstitions seem respectable. West gave his low laugh, apparently signifying agreement.

The psychologist, meanwhile, was waiting for West to go to sleep. He had not done so for more than a few hours altogether. By the third day, when they were sailing in sunshine with a gentle breeze, the psychologist expected to see signs of exhaustion appear in West. Instead, West put on his bathing suit and took a long vigorous swim beside the boat.

Back at a restaurant near Portland before they'd gone out into the storm, while they'd been sharing the meal that most of them soon regretted, West had told them, "I build computers." Although he spoke at some length about certain extraordinary sounding, new computing systems, the others came away uncertain about what role, if any, he had played in their construction. They felt only that whatever he did for a living, it was probably interesting and obviously important.

One time while West was manning the tiller, the psychologist asked him how he had learned to sail. West didn't answer. A little later on, thinking he hadn't heard the question, the psychologist inquired again.

"You already asked me that," West snapped. After a moment's silence, he wet his lips and explained that he had taught himself mostly, as a boy.

On another occasion, just to make conversation, one of the crew asked West what sort of computer he was building now. West made a face and looked away, and muttered something about how *that* was work and this was his vacation and he would rather not think about *that*.

The people who shared the journey remembered West. The following winter, describing the nasty northeaster over dinner, the captain remarked, "That fellow West is a good man in a storm." The psychologist did not see West again, but remained curious about him. "He didn't sleep for four nights! *Four whole nights.*" And if that trip had been his idea of a vacation, where, the psychologist wanted to know, did he work?

1

HOW TO MAKE
A LOT OF MONEY

FOR A TIME after the first pieces of Route 495 were laid down
across central Massachusetts, in the middle 1960s, the main haz-
ard to drivers was deer. About fifteen years later, although traffic
went by in processions, stretches of the highway's banks still
looked lonesome. Driving down 495, you passed some modern
buildings, but they quickly disappeared and then for a while there
would be little to see except the odd farmhouse and acres of trees.
The highway traverses some of the ghost country of rural Massachusetts. Like Troy, this region contains evidence of successive
sackings: in the pine and hardwood forests, which now comprise
two-thirds of the state, many cellar holes and overgrown stone
walls that farmers left behind when they went west; riverside tex-
tile mills, still the largest buildings in many little towns, but their
windows broken now, their machinery crumbling to rust and the
business gone to Asia and down south. However, on many of the
roads that lead back behind the highway's scenery stand not
woods and relics, but brand-new neighborhoods, apartment
houses, and shopping centers. The roads around them fill up with
cars before nine and after five. They are going to and from commercial buildings that wear on their doors and walls descriptions
of new enterprise. Digital Equipment, Data General — there on

the edge of the woods, those names seemed like prophecies to me, before I realized that the new order they implied had arrived already.

A few miles north of the junction of Route 495 and the Massachusetts Turnpike, off an access road, sits a two-story brick building, surrounded by parking lots. A sign warns against leaving a car there without authority. The building itself looks like a fort. It has narrow windows, an American flag on a pole out front, a dish antenna on a latticed tower. Mounted on several corners of the roofs, and slowly turning, are little TV cameras.

This is Building 14A/B — 14B was fastened seamlessly to 14A. Some employees call the place "Webo," but most refer to it as "Westborough," after the name of the town inside whose borders the building happens to exist. "Westborough" is worldwide headquarters of the Data General Corporation. Driving up to the building one day with one of the company's public relations men, I asked, "Who was the architect?"

"We didn't have one!" cried the beaming press agent.

Company engineers helped to design Westborough, and they made it functional and cheap. One contractor who did some work for Data General was quoted in *Fortune* as saying, "What they call tough auditing, we call thievery." However they accomplished it, Westborough cost only about nineteen dollars a square foot at a time when the average commercial building in Massachusetts was going for something like thirty-four dollars a foot. But looks do matter here. The company designed Westborough not just for the sake of thriftiness, but also to make plain to investors and financial analysts that Data General really is a thrifty outfit. "There's no reason in our business to have an ostentatious display," a company analyst for investor relations explained. "In fact, it's detrimental."

The TV cameras on the roofs, the first defense against unscrupulous competitors and other sorts of spies and thieves, must comfort those who have a stake in what goes on inside. As for me, I imagined that somewhere in the building men in uniforms were watching me arrive, and I felt discouraged from walking on the grass.

The only door that opens for outsiders leads to the front lobby. A receptionist asks you to sign a logbook, which inquires if you are an American citizen, wants your license plate number, and so on. Still you cannot pass the desk and enter the hallways beyond — not until the employee you want to see comes out and gives you escort. When I inquired, the cheerful young receptionist said that once in a great while some outsider would *try* to break the rules and *try* to slip inside.

The lobby could belong to a motor inn. It has orange carpeting and some chairs and a sofa upholstered in vinyl, on which salesmen and would-be employees languish, awaiting appointments. Now and then, a visitor will stand and gaze into a plastic case. It contains the bare bones of a story that will feed the dreams of any ambitious businessman. THE FIRST NOVA, reads a legend on the case. Inside sits a small computer, about the size of a suitcase, with a cathode-ray tube — a thing like a television screen — beside it. A swatch of prose on the back wall, inside the case, explains that this was the first computer that Data General ever sold. But the animal in there isn't stuffed; the computer is functioning, lights on it softly blinking as it produces on the screen beside it a series of graphs — ten years' worth of annual reports, a précis of Data General Corporation's financial history.

Left to their own devices, the engineers who worked in the basement of Building 14A/B could surely have produced a flashier display, but a visitor from Wall Street who had never paid attention to this company before might have felt faint before the thing. The TV screen was blue. The graphs, etched in white, appeared in rotating sequence, and each one bore a name. "Cumulative Computers Shipped Since Our Founding" started with 100 in 1969 and went right up to 70,700 in 1979. The image vanished. "Net Sales" appeared, to show that revenues had ascended without a hitch from nothing in 1968 to $507.5 million in 1979. That graph went away and in its place came one describing profit margins. These hardly varied. The profits just rolled in, year after year, along a nearly straight line, at about 20 percent (before taxes) of those burgeoning net sales.

Someone unaccustomed to reading financial reports might have missed the full import of the numbers on the screen, the glee and madness in them. But anyone could see that they started small and got big fast. Mechanically, monotonously, the computer in the case was telling an old familiar story — the international, materialistic fairy tale come true.

The first modern computers arrived in the late 1940s, and although many more or less single-handed contributions fostered the technology, they did so mainly in the shade of a familiar association in America among the military, universities and corporations. On the commercial side, IBM quickly established worldwide hegemony; it brought to computers the world's best sales force, all dressed in white shirts and blue suits. For some years the computer industry consisted almost exclusively of IBM and several smaller companies — "IBM and the seven dwarfs," business writers liked to say. Then in the 1960s IBM produced a family of new computers, called the 360 line. It was a daring corporate undertaking. "We're betting the company," one IBM executive remarked. Indeed, the project cost somewhat more than the development of the atom bomb, but it paid off handsomely. It guaranteed for a long time to come IBM's continued preeminence in the making of computers for profit. Meanwhile, though, new parts of the business were growing up, and out from under IBM.

In the early days, computers inspired widespread awe and the popular press dubbed them giant brains. In fact, the computer's power resembled that of a bulldozer; it did not harness subtlety, though subtlety went into its design. It did mainly bookkeeping and math, by rote procedures, and it did them far more quickly than they had ever been done before. But computers were relatively scarce, and they were large and very expensive. Typically, one big machine served an entire organization. Often it lay behind a plate glass window, people in white gowns attending it, and those who wished to use it did so through intermediaries. Users were like supplicants. The process could be annoying. Scientists and engineers, it seems, were the first to express a desire for a relatively inexpensive computer that they could operate themselves. The result was a machine called a minicomputer. In time, the demand for such a machine turned out to be enormous. Probably IBM could not have controlled this new market, the way it did the one for large computers. As it happened, IBM ignored it, and so the field was left open for aspiring entrepreneurs — often, in this case, young computer engineers who left corporate armies with dreams of building corporate armies of their own.

For many years sociologists and others have written of a computer revolution, impending or in progress. Some enthusiasts have declared that the small inexpensive computer inaugurated a new phase of this upheaval, which would make computers instruments of egalitarianism. By the late seventies, practically every organization in America had come to rely upon computers, and ordinary citizens were buying them for their homes. Within some organizations small bands of professionals had exercised absolute authority over computing, and the proliferation of small computers did weaken their positions. But in the main, computers altered techniques and not intentions and in many cases served to increase the power of executives on top and to prop up venerable institutions. A more likely place to look for radical change was inside the industry actually producing computers. Generally, that industry grew very big and lively, largely because of a single invention.

Shortly after World War II, decades of investigation into the internal workings of the solids yielded a new piece of electronic hardware called a transistor (for its actual invention, three scientists at Bell Laboratories won the Nobel Prize). Transistors, a family of devices, alter and control the flow of electricity in circuits; one standard rough analogy compares their action to that of faucets controlling the flow of water in pipes. Other devices then in existence could do the same work, but transistors are superior. They are solid. They have no cogs and wheels, no separate pieces to be soldered together; it is as if they are stones performing useful work. They are durable, take almost no time to start working, and don't consume much power. Moreover, as physicists and engineers discovered, they could be made very small, indeed microscopic, and they could be produced cheaply in large quantities.

The second crucial stage in the development of the new electronics came when techniques were developed to hook many transistors together into complicated circuits — into little packets called integrated circuits, or chips (imagine the wiring diagram of an office building, inscribed on the nail of your little toe). The semiconductor industry, which is named for the class of solids out of which transistors are made, grew up around these devices and began producing chips in huge quantities. Chips made spaceships and pocket calculators possible. They became the basic building blocks of TVs, radios, stereos, watches, and they made computers ubiquitous and varied. They did not eliminate the sizable, expensive computer; they made it possible for the likes of IBM to produce machines of increased speed and capability and still make handsome profits without raising prices much. At the same time, the development of chips fostered an immense and rapid growth of other kinds of computing machines.

After mainframes, as the big computers were known, came the cheaper and less powerful minicomputers. Then the semiconductor firms contributed the microprocessor, the central works of a computer executed on a chip. For a while, the three classifications really did describe a company's products and define its markets, but then mainframers and microcomputer companies started making minis and minicomputer companies added micros and things that looked like mainframes to their product lines. Meanwhile, a host of frankly imitative enterprises started making computers and gear for computers that could be plugged right into systems built around the wares of the big successful companies.

These outfits went by the names of "plug compatibles" and "third-party peripheral manufacturers"; those who lost some business to them called them "knockoff companies." Probably they helped maintain competition in prices. Many "software" houses sprang up, to write programs that would make all those computers actually do work. Many customers, such as the Department of Defense, wanted to buy complete systems, all put together and ready to run with the turn of a key; hence the rise of companies known as original equipment manufacturers, or OEMs — they'd buy gear from various companies and put it together in packages. Some firms made computer systems for hospitals; some specialized in graphics — computers that draw pictures — and others worked on making robots. It became apparent that communications and computing served each other so intimately that they might actually become the same thing; IBM bought a share in a satellite, and that other nation-state, AT&T, the phone company, started making machines that looked suspiciously like computers. Conglomerates, of which Exxon was only the largest, seemed determined to buy up every small computer firm they could. As for those who observed the activity, they constituted an industry in themselves. Trade publications flourished; they bore names such as *Datamation, Electronic News, Byte, Computermania*. IBM, one executive of a mainframe company once said, represented not competition but "the environment," and on Wall Street and elsewhere some people made a business solely out of attempting to predict what the environment would do next.

I once asked a press agent for a computer company what was the reason for all this enthusiasm. He held a hand before my face and rubbed his thumb across his fingers. "Money," he whispered solemnly. "There's so goddamn much money to be made." Examples of spectacular success abounded. The industry saw some classic dirty deals and some notable failures, too. RCA and Xerox lost about a billion dollars apiece and GE about half a billion making computers. It was a gold rush. IBM set up two main divisions, each one representing the other's main competition. Other companies did not have to invent competitors and did somewhat more of their contending externally. Some did sometimes use illicit tools. Currying favor, seeking big orders for chips, some salesmen of semiconductors, for instance, were known for whispering to one computer maker news about another computer maker's latest unannounced product. Firms fought over patents, marketing practices and employees, and once in a while someone would get caught stealing blueprints or other documents, and for these and other reasons computer companies often went to court. IBM virtually resided there. Everyone sued IBM, it seemed. The biggest suit, the *Jarndyce* v. *Jarndyce* of the industry, involved the Justice Department's attempt to break up IBM. Virtually an entire large law firm was created to defend IBM in this case, which by 1980 had run ten years and had been in continuous trial for several.

Data General took its place in this bellicose land of opportunity in 1968, as a "minicomputer company." By the end of 1978 this increasingly undescriptive term could in some senses be applied to about fifty companies. Their principal but by no means their only business, the manufacture and sale of small computers, had grown spectacularly — from about $150 million worth of shipments in 1968, to about $3.5 billion worth by 1978 — and it would continue to grow, most interested parties believed, at the rate of about 30 percent a year. By 1978 Data General ranked third in sales of minicomputers and stood among the powers in this segment of the industry. The leader was Digital Equipment Corporation, or DEC, as it is usually called. DEC produced some of the first minicomputers, back in the early sixties. Data General was the son, emphatically the son, of DEC.

A chapter of DEC's official history, a technical work that the company published, describes the making of a computer called the PDP-8. DEC sent this machine to market in 1965. It was a hit. It made DEC's first fortune. The PDP-8, says the official history, "established the concept of minicomputers, leading the way to a multibillion dollar industry." But the book doesn't say that Edson de Castro, then an engineer in his twenties, led the team that designed the PDP-8. The technical history mentions de Castro only once, briefly, and in another context. They expunged de Castro. In 1968 de Castro and two other young engineers seceded from DEC. Several completely different versions of their flight exist and have by now acquired the impenetrable quality of myth. Did they quit because, after long and heartfelt labor on a new design, they found that DEC's management would not build their new machine? DEC's management did turn down a new design of de Castro's, and afterward, along with a man from another company named Herb Richman, de Castro and the two other engineers from Digital incorporated Data General and started building their own minicomputer. But did they design this new machine after they seceded, or had they done that job in secret, using DEC's facilities, while still on DEC's payroll? One version of the story suggested the latter. More than ten years later, DEC's founder and president would tell reporters from *Fortune,* "What they did was so bad we're still upset about it." But DEC never sued Data General's founders, and clearly there were other reasons why Digital might have become upset. For within a year, de Castro and company had set up shop in DEC's own territory and had started raking in the loot.

They rented space in what had been a beauty parlor, in the former mill town of Hudson, Massachusetts. Practically all that remains of that time is a black-and-white photograph of this first headquarters. In the foreground stand four young men with short hair, wearing white shirts and skinny ties and the sort of plain black shoes that J. Edgar Hoover's men favored. They are engaged in what is obviously meant to look like routine conversation. The linoleum floor, the metal furniture, evoke motor vehicle departments, and the youths in the picture could be members of some junior chamber of commerce, playing capitalists for a day. Not shown in this bemusing picture is the shrewd and somewhat older lawyer from a large New York firm who helped Data General's founders raise their capital and who became a crucial member of their team. What also doesn't show is the fact that some of

these young men were already computer engineers of no mean repute — their age in this case was no impediment, for computer engineers like athletes often blossom early.
They started Data General at an auspicious time. In the late
1960s, the period memorialized in John Brooks's *The Go-Go
Years,* venture capital (among other things) abounded, and although they started out with only $800,000, more lay in reserve.
They also entered a good territory for fledglings. They could not
have dreamed of moving in on IBM's markets without truly vast
amounts of capital. But the people who bought minicomputers —
engineers, scientists, and, mainly, purchasing agents of OEMs —
understood the machines. A new manufacturer could reach them
through relatively inexpensive ads in the trade journals, and
didn't need to build a service organization right away, since these
customers could take care of themselves. These were also the sorts
of customers who could be expected to embrace a newcomer, if
the price was right; they'd prefer a bargain to a brand name.
But around the time when Data General established itself in
the beauty parlor, other entrepreneurs were starting up minicomputer companies at the rate of about one every three days.
Only a few of those other new outfits survived the decade, whereas
Data General, before it had exhausted its first and fairly modest
dose of capital, achieved and never fell from that state of grace, a
positive cash flow. Why?
The company's first machine, the NOVA, had a simple elegance
about it that computer engineers I've talked to consider admira-
ble, for its time. It had features that DEC's comparable offering
didn't share, and it incorporated the latest, though not fully
proven, advances in chips. Data General could build the NOVA
very cheaply. Such an important advantage can depend, in computers, on small things. In the case of the NOVA, the especially
large size of the printed-circuit boards — the plates on which the
chips are laid down — made a crucial difference. For several reasons, large boards tend to reduce the amount of hardware in a
computer. Data General used boards much larger than the ones

that DEC was using. Speaking of this difference and of other less important ones, one engineer remarked, "The NOVA was a triumph of packaging."

Good machines don't guarantee success, though, as RCA and Xerox and others had discovered. Herb Richman, who had helped to found Data General, said, "We did *everything* well." Obviously, they did not manage every side of their business better than everyone else, but these young men (all equipped with large egos, as one who was around them at this time remarked) somehow managed to realize that they had to attend with equal care to all sides of their operation — to the selling of their machine as well as to its design, for instance. That may seem an elementary rule for making money in a business, but it is one that is easier to state than to obey. Some notion of how shrewd they could be is perhaps revealed in the fact that they never tried to hoard a majority of the stock, but used it instead as a tool for growth. Many young entrepreneurs, confusing ownership with control, can't bring themselves to do this.

When they chose their lawyer, who would deal with the financial community for them, they insisted that he invest some of his own money in their company. "We don't want you running away if we get in trouble. We want you there protecting your own money," Richman remembered saying. Such an arrangement, though not illegal, might raise some eyebrows in some corners of the Bar Association. But the lawyer said, again according to Richman, "That's the first time anyone made an intelligent proposition to me." Richman also remembered that before they entered into negotiations over their second public offering of stock, after the company had been making money for a while and the stock they'd already issued had done very well indeed, their lawyer insisted that each of the founders sell some of their holdings in the company and each "take down a million bucks." This so that they could negotiate without the dread of losing everything ("Having to go back to your father's gas station," Richman called that nightmare). As for the name of the theory behind selling enough stock to become millionaires, Richman told me, "I don't know how you put it in the vernacular. We called it the Fuck You Theory."

In the computer business, your market can be your fate. Although by the late 1970s it was hard to define a company's place in the industry by the sorts of machines that it made, certain broad historic distinctions in ways of doing business still divided a large part of the industry into three segments. The differences showed up in the nature of a company's expenditures. IBM and other mainframe companies spent more money selling their products and serving their customers than they did in actually building their machines. They sold their computers to people who were actually going to use them, not to middlemen, and this market required good manners. Microcomputer companies sold equipment as if it were corn, in large quantities; they spent most of their money making things and competed not by being polite but by being aggressive. Minicomputer companies split the differences more or less; they sold some machines and service to actual users, but spent most of their money on hardware and did a big business by selling machines in quantity to OEMs.

From these distinctions, others hung. A seasoned executive in marketing explained, "With micros it's even more competitive, but historically the world of minicomputers is very rough-and-tumble. IBM would say, 'You got a problem, Mr. Customer? A team of four will be there in an hour.' Implicitly a Data General would say to its customers, 'You have to look out for yourselves.' The sophisticated customer, particularly the OEM who buys a lot of computers and looks for discounts, not service, goes for minis. They're capable of living in a rough-and-tumble world. And I'm not sure that IBM, with its organization, can compete in the traditional minicomputer market. It's like putting a goldfish in a bowl with a piranha."

So you could say that Data General entered a territory that

asked for a certain brashness. And you could also say that life in this territory became less decorous than it had been, when Data General came along. They set out to get noticed, first of all. In the lobby at Westborough hangs a copy of the first advertisement that Data General ever ran. It consists of just one page. On one side of the page is a grainy photograph of a man's face. This person looks about to do something very mean. On the other side of the ad, he speaks: "I'm Ed de Castro, president of Data General Corporation. Seven months ago we started the richest small computer company in history. This month we're announcing our first product: the best small computer in the world." The message goes on for a while and winds up as follows:

> *Because if you're going to make a small inexpensive computer you have to sell a lot of them to make a lot of money. And we intend to make a lot of money.*

This ad's chief architect, a man named Allen Kluchman, who was the company's first director of marketing, told me with a smile, "That ad was independent of any aspect of Mr. de Castro's personality that I knew about at that time. He's the shyest guy I know. He's essentially a pretty humble, private guy."

The ad achieved a certain local fame. It said what many others presumably were thinking, but what none of them felt they should say publicly. For some years thereafter, most of Data General's advertisements contained something slightly brazen. One of the best-known ads wasn't published — some people in the company were by then apparently having second thoughts about the firm's image. But a copy of this unpublished ad hangs in de Castro's office. Over the Data General logo, on a field of white and blue, it reads:

> *They Say IBM's Entry Into Minicomputers Will Legitimize The Market. The Bastards Say, Welcome.*

Before Data General unveiled the NOVA in 1969 — at the industry's yearly fair, the National Computer Conference — the marketeer Kluchman talked a trade magazine into putting a picture of the NOVA on its cover. They rented billboards on the road from the airport to the conference and put a picture of the NOVA on them; at the hotel where most of the people attending the conference stayed, they talked the management into having bellboys distribute free copies of the *Wall Street Journal* with Data General's advertising flyer inside; at the show itself, they raised the placard bearing their company's name higher than any other. When it came to pricing their machine, they announced extraordinary discounts for customers who bought machines in quantity. Never mind that customers had to buy a virtual warehouse of NOVAs to get the truly big discount. Data General had brought a new ferocity, a bit of Forty-second Street, to the pricing of minicomputers.

"DEC owned 85 percent of the business and there was no strong number two. We had to distinguish ourselves from DEC," Kluchman remembered. "DEC was known as a bland entity. Data General was gonna be unbland, aggressive, hustling, offering you more for your money. . . . We spread the idea that Data General's salesmen were more aggressive than DEC's, and they were, because ours worked on commissions and theirs worked on salaries. But I exaggerated the aggressiveness."

According to Kluchman, DEC actually gave them some help in setting up "the Hertz-Avis thing." DEC's management, he said, ordered their salesmen to warn their customers against Data General. "It was great! Because their customers hadn't heard about us." Kluchman imagined DEC's salesmen telling DEC's customers that a dangerous new company was on the prowl, and DEC's customers responding to this news by saying, "Where is this Data General, so we can be sure to avoid them? What's Data General's phone number, so we'll be sure not to call it?" Kluchman laughed. "The calls just *rolled* in. DEC's customers would say, 'We hear you're bad guys. You must be doing something we oughta know about.' "

And thinking back to those first heady days, when nearly every little strategy seemed to pay off, and the first millions started coming in, Kluchman said, "It was probably more fun than I or anybody else has ever had in business. It was great ego satisfaction. It was just a *pure gas.*"

At the end of fiscal 1978, after just ten years of existence, Data General's name appeared on the list of the nation's five hundred largest industrial corporations — in that band of giants known as the Fortune 500. It stood in five-hundredth place in total revenues, but much higher in respect to the various indices of profit, and for a while climbed steadily higher on the list. Surely by 1980 such a record entitled Data General to respectability. But some trade journalists still looked askance at the company; one told me Data General was widely known among his colleagues as "the Darth Vader of the computer industry." Investors still seemed jittery about Data General's stock. An article published in *Fortune* in 1979 had labeled Data General "the upstarts," while calling DEC "the gentlemen." The memory of that article, particularly the part that made it sound as if Data General routinely cheated its customers, still rankled Herb Richman.

Building 14A/B is essentially divided into an upstairs and a downstairs, and in one corner of the upstairs the corporate officers reside. A wall of glass separates them from the rest of the company. There is no mahogany here. If there is ostentation in the bosses' quarters, it is ostentation in reverse. The table in their conference room, it was proudly said, was the same that they had used when the company was small. Richman's office was comparatively plush. But saying, "We consider ourselves the Robert Hall of the computer industry," Richman pointed out that he had paid for all his furnishings himself and that what looked like paneling on his walls was really just wallpaper.

Among the founders of the company, only de Castro — the

much-talked-about president — and Richman remained engaged in daily operations. Richman had come up through the industry in sales — a supersalesman, some called him — and he had created and run Data General's sales force, which was known if not notorious for its aggressiveness. Curly-haired, trim and in his forties, Richman wore a nicely tailored denim jacket and no tie. "I'm one of the few guys that money made a nice guy out of," he said. "Before, I was just driven, clawing. . . . Success has made me more rational and introspective." He remarked that not long ago he had been playing tennis with a man who had seemed to him just an ordinary fellow, but then he had found out that the man was actually president of an oil company. "And it was one of the largest oil companies in the world, and I was just in awe of him," said Richman. He added, softly, "And yet I bet my net worth greatly exceeded his."

The stock that Richman himself owned in Data General was worth about $13 million then, but, he seemed to say, he was unhappy with the way certain organs of the press depicted his company's achievements. They were, Richman believed, too often depicted as "ruffians," not as merely rough, which they were proud to be. "We agree that a lot of things we've done around here are wild," he said. "But we can't understand why we're tabloid, instead of the *New York Times*."

Some part of Data General's reputation was easy to explain. The company had promoted it themselves, and maybe it had gotten a little out of hand. Richman suggested, "We've done so much so well for so long that everyone seems to think we have to be doing something illegal." A good point, but not a full accounting.

Some years back, in the early seventies, a company called Keronix accused Data General's officers of arranging the burning of a factory. Keronix had been making computers that performed almost identically to Data General machines. The theory was that Data General had taken a shortcut in attempting to get rid of this competitor. In time, the courts found no basis for those charges and dismissed them. Indeed, it seemed preposterous to think that the suddenly wealthy executives of Data General would risk everything, including jail, and resort to arson, just to drive away what was, after all, a small competitor. But Wall Street didn't see it that way, apparently. When Keronix made its accusation, Data General's stock plummeted; there was such a rush to unload it that the New York Exchange had to suspend trading in it for a while. More peculiar was the fact that many years later, some veteran employees, fairly far down in the hierarchy, would say privately that they believed someone connected with the company had something to do with that fire. Not the officers, but some renegade within the organization. They had no basis for saying so, no piece of long-hidden evidence. It seemed to me that this was something that they wanted to believe.

I got this feeling more than once. Turning down the road to Building 14A/B one day, a veteran engineer pointed out the sign that warned against unauthorized parking. "The first sign you see says Don't," he remarked. He imagined another sign by the road; it would say: Use of Excessive Force Has Been Approved. The engineer laughed and laughed at the thought.

In a land of tough and ready companies, theirs, some of Data General's employees seemed to want to think, was the toughest and the readiest around.

Certainly Data General's reputation had other underpinnings besides advertisements and imagination. In an industry where sharp marketing practices were common, Data General's were as sharp as any, and by the late 1970s competitors were challenging some of them in federal court. To the contention, leveled in *Fortune,* that Data General played especially rough with its customers, it was only fair to add that many of Data General's customers knew very well what sort of market they were in, and moreover, it was clear that the company could not have survived if most of its customers had not felt at least fairly satisfied. But Data General was litigious, toward customers as well as others. "Sure," said Richman, "if people don't pay us or breach our contract, we litigate 'em." They did so at least in part to assure Wall Street that they weren't the sort of company that would accumulate a crippling number of bad debts.

The salient feature of Data General, however — what that sharp-eyed, astonished visitor from Wall Street would have pondered — was its growth. This was indeed the industry's salient characteristic. In the main, computer companies that were not dying were growing; they had to do so just to stay alive, it seemed. But no company whose primary business was making computers had grown more rapidly than Data General. Bursts of growth were not uncommon, but Data General had been bursting for a decade, and what's more, it had been maintaining the highest profit margins in the industry next to IBM's. All this would have impressed the analyst from Wall Street, of course, but would also have given him pause.

Building 14A/B and its sparse furnishings, the facts that Data General paid its stockholders no dividends and that its top managers dispensed to themselves and other officers exceptionally small salaries, meting out rewards in the form of stock instead — all were signs of a common purpose. The company had displayed extravagance when it came to financing its tendency to go to court. Otherwise, the management seemed bent on saving all their cash to feed the hungry beast of growth. And, of course, the more this beast gets fed, the bigger it becomes, the more it wants to eat. It is one thing for a company with revenues of a million dollars a year to grow 30 or 40 percent in a year and quite another for a half-a-billion-dollar company to pull off the same trick.

Analysts on Wall Street sometimes become boosters of the companies they follow. Looking for an opinion that was certain to be disinterested, I asked an old friend, a veteran analyst of securities, to take a look at Data General's numbers. He had the advantage of never having followed the company before, and in return for anonymity he agreed to my proposal. A couple of weeks later, he called me back. It seemed to him that Data General was bent on continuing to grow at 30 to 40 percent a year. He pointed out that this meant large growth in everything — in the need for capital, new buildings, new employees. Between 1974 and 1978, for instance, Data General had hired about 7,000 new employees, roughly tripling its numbers; in one year alone the company had increased its ranks by 71 percent. The analyst imagined the difficulties of finding that many qualified people so quickly. And what must it be like, he asked, to work at a place like that? You'd come to work some morning and suddenly find yourself in charge of a dozen new people, or suddenly beneath a new boss to whom you would have to prove yourself all over again. "That sort of growth puts a strain on everything," the analyst concluded. "It's gonna be intriguing to see if they get caught." He thanked me for putting him onto such a marvelous entertainment.

Where did the risks lie? Where could a company go badly wrong? In many cases, a small and daily growing computer company did not fall on hard times because people suddenly stopped wanting to buy its products. On the contrary, a company was more likely to asphyxiate on its own success. Demand for its products would be soaring, and the owners would be drawing up optimistic five-year plans, when all of a sudden something would go wrong with their system of production. They wouldn't be able to produce the machines that they had promised to deliver. Lawsuits might follow. At the least, expensive parts would sit in inventory, revenues would fall, customers would go elsewhere or out of business themselves. Data General got one leg caught in this trap back around 1973. Six years later, a middle-level executive, sitting in an office upstairs at headquarters, remembered that time: "We were missing our commitments to customers. We just grossly fucked over our customers. We actually put some entrepreneurs out of business and I think some of them may have lost their houses. But we recovered from our shipment problems and never repeated them."

Another way of fouling up had less to do with a company's own growth than with the growth occurring all around it. From observers of the industry came such comments as: "Things change fast in the computer business. A year is a hell of a long time. It's like a year in a dog's life." In every segment of the industry, companies announced small new products for sale every day. Companies brought out new lines of computers, much more powerful than the ones they replaced, only every few years or so; but considering all the work that went into them and the fact that they required a redirection of effort throughout a company, the pace at which these major announcements came was very rapid too. Conventional wisdom held that if a company fell very far behind its competitors in producing the latest sorts of machines, it would have a hard time catching up. And failure to stay abreast could have serious consequences, because major new computers played crucial roles in the other business of the companies; they helped them sell all their little products and, often, their older types of machines.

At some companies the task of guarding against this sort of crisis fell mainly to engineers, working below decks, as it were. Executives might make the final decisions about what would be produced, but engineers would provide most of the ideas for new products. After all, engineers were the people who really knew the state of the art and who were therefore best equipped to prophesy changes in it. At Data General, an engineer could play such an important role. It was there for the taking. The president, de Castro, liked "self-starters," it was said. Initiative was welcomed at Data General, and in the late seventies it appeared that the company had need of some initiative from its engineers. For Data General was in a bind. The firm had fallen behind the competition: it hadn't yet produced the latest big thing in minicomputers.

Early in 1979 the businessman who told me about Data General's problems and recovery back in '73 hit upon a heroic metaphor for success in the computer business. "The major thing," he said, "is avoiding the big mistake. It's like downhill ski racing: Can you stay right on that edge beside disaster? At Data General we keep coming up with these things that are basically acts of recovery. What Tom West and his people are doing is a great act of recovery."

2

THE WARS

TOM WEST went to work these days in freshly laundered blue jeans or pressed khakis, in leather moccasins and in solid-colored long-sleeved shirts, with the sleeves rolled up in precise folds, like the pages of a letter, well above his bony elbows. He expostulated with his hands. When dismissing someone or some idea or both, he made a fist and then exploded it, fingers splaying wide. The gesture was well known to the engineers who worked for him. Long index fingers inserted under either side of the bridge of his glasses signified thought, and accompanied by a long "Ummmmmmmmh" warned that some emphatic statement was near. He kept his car and his office as neat as the folds in his sleeves. He was decisive, and in his manner, exact. For all of that, he was vague. "When I first went to work, he was my boss," an engineer said of West, "and it was amazing! Half the time I couldn't figure out what he was saying." He was not always this way, said one of his oldest colleagues, but sometimes you got the feeling that West expected you to be on his secret wavelength, and if you weren't he'd be disappointed in you. If you weren't, that was your problem. He didn't have time to explain.

Seen at the wheel of his sporty red Saab, driving to work down 495, West made a picture of impatience. His jaw was set, he had a forward lean. Sometimes he briefly wore a mysterious smile. He was a man on a mission.

Into the world of the minicomputer a new thing had been born, a class of computer known as a 32-bit supermini. West said, with characteristic enthusiasm: "Everyone thinks they want one now. It's an emotional issue. In fact, it's kind of a fire storm." As for the present state of affairs, sometimes he called it "a disaster." Sometimes he would say, "We're gonna get schmeared if we don't react to VAX."

A number of Data General's rivals had produced 32-bit superminis, and the most important from West's point of view was the computer that DEC had recently sent to market, a machine called the VAX 11/780. Data General, meanwhile, had not yet produced a computer of this class. Many people, including West, believed that they must do so, and in fairly short order. Partly it was a matter of keeping up appearances: customers get married to their computer companies in many different ways and they don't usually want to get or stay married to a company that has fallen behind the state of the art. Besides, you had to grab a piece of the new market for the 32-bit supermini because that market was huge and growing fast; most observers agreed that it would be worth several billion dollars by the middle 1980s. You did not have to be the first company to produce the new kind of machine; sometimes, in fact, it was better not to be the first. But you had to produce yours before the new market really opened up and customers had made other marriages. For once they are lost, both old and prospective customers are often gone for good.

It had been painful for West and for a number of engineers working with him at Westborough to watch DEC's VAX go to market, to hear it described as "a breakthrough," and not have a brand-new machine of their own to show off. It had been painful for them to read in the trade press of the VAX's growing success; VAX was beginning to look like one of those best-sellers that come along only once in a while. But by the fall of 1978 West had drawn around him a team of enthusiastic engineers and they were finally working on their own supermini, which they had nick-named Eagle. A new computer, especially one of this class, does not get built in a month. Often it takes years to bring one to life. But it wasn't too late, West was saying. Not if they could build this computer in the record time of something like a year. This was at last "the right machine." This was "the answer to VAX." At times West fancied that this computer would become the source of Data General's continued ascent in the Fortune 500. "This is the second billion," he said. His doubts he did not share widely.

Secretly, West felt afraid of VAX. DEC had published a great deal of technical literature describing VAX, and West had read all of it. Nothing in this material had made him feel that his team's approach was inferior to DEC's. In some engineers, however, reading does not constitute knowing. For them, touch is the first of the senses. And so, one holiday morning in 1978, when his team was already well launched on the building of its own machine, West went away from Westborough to have a look at a VAX for himself.

He traveled to a city, which was located, he would only say, somewhere in America. He walked into a building, just as though he belonged there, went down a hallway, and let himself quietly into a windowless room. The floor was torn up; a sort of trench filled with fat power cables traversed it. Along the far wall, at the end of the trench, stood a brand-new example of DEC's VAX, enclosed in several large cabinets that vaguely resembled refriger-ators. But to West's surprise, one of the cabinets stood open and a man with tools was standing in front of it. A technician from DEC, still installing the machine, West figured.

Although West's purposes were not illegal, they were sly, and he had no intention of embarrassing the friend who had given him permission to visit this room. If the technician had asked West to identify himself, West would not have lied, and he wouldn't have answered the question either. But the moment went by. The technician didn't inquire. West stood around and watched him work, and in a little while, the technician packed up his tools and left.

Then West closed the door, went back across the room to the computer, which was now all but fully assembled, and began to take it apart.

The cabinet he opened contained the VAX's Central Processing Unit, known as the CPU — the heart of the physical machine. In the VAX, twenty-seven printed-circuit boards, arranged like books on a shelf, made up this thing of things. West spent most of the rest of the morning pulling out boards; he'd examine each one, then put it back.

Across the surfaces of a typical computer's printed-circuit boards stand columns of small rectangular boxes, with metal legs descending from their sides. They might be some odd strain of caterpillar bred for mathematical ability. In fact, each of these boxes holds inside it another box, as it were — the intricate integrated circuitry known as a chip. Etched into the boards, among the housings of the chips, run many silvery bands; they make patterns like the tracks in large railroad yards.

Some boards are colorful and most finished ones please the eye. A computer's boards seem to show order triumphing in complexity. They look as if they make sense, but not in the way the moving parts of an engine make sense. The form on the surface of a board does not imply its function. It's difficult but possible to get inside the littlest boxes inside the boxes that constitute a modern computer, and bringing back the details, to create a functionally equivalent copy of the machine. Reverse engineering is the name for that art.

West called it "knockoff copy work." He had a cleaner, simpler purpose. He examined the outside of the VAX's chips — some had numbers on them that were like familiar names to him — and he counted the various types and the quantities of each. Later on, he looked at other pieces of the machine. He identified them generally too. He did more counting. And when he was all done, he added everything together and decided that it probably cost $22,500 to manufacture the essential hardware that comprised a VAX (which DEC was selling for somewhat more than $100,000). He left the machine exactly as he had found it.

"I'd been living in fear of VAX for a year," West said afterward, while driving along 495 one evening. "I wasn't really into G-Two. VAX was in the public domain and I wanted to see how bad the damage was. I think I got a high when I looked at it and saw how complex and expensive it was. It made me feel good about some of the decisions we've made."

Looking into the VAX, West had imagined he saw a diagram of DEC's corporate organization. He felt that VAX was too complicated. He did not like, for instance, the system by which various parts of the machine communicated with each other; for his taste, there was too much protocol involved. He decided that VAX embodied flaws in DEC's corporate organization. The machine expressed that phenomenally successful company's cautious, bureaucratic style. Was this true? West said it didn't matter, it was a useful theory. Then he rephrased his opinions. "With VAX, DEC was trying to minimize the risk," he said, as he swerved around another car. Grinning, he went on: "We're trying to maximize the win, and make Eagle go as fast as a raped ape."

Some of the engineers closest to West suspected that if he weren't given a crisis to deal with once in a while, he would create one. To them he seemed so confident and happy in an emergency. But as for this big crisis in the little world of Westborough engineering, although West had made it his own, no one could say he had invented it.

Why had Data General failed to produce a rival to VAX? When trade journalists asked this question, Data General put on a bold face of course, and suggested in essence that all was proceeding according to plan. In fact, years before the appearance of VAX, Data General engineers had foreseen the advent of such machines. For about five years they had been trying to produce one of their own. But there had been problems. They had made some false starts, and the engineers involved had been arguing over who was going to produce this new machine and what it would be.

Some computer engineers harbor strong feelings toward their new designs, like Cossacks toward their horses. Carl Alsing, a veteran engineer and one of West's cadre, told the fable of an engineer, who, upon being informed that his plans for a new machine had been scrapped by the managers of his company, got a gun and murdered a colleague whose design had been accepted. Alsing said he thought that such a murder really happened but that a woman was probably involved — yet it came, he said, to much the same thing.

The history of what West and some of his associates called their "wars" began in the mid-seventies. Data General had gotten over its manufacturing problems; the fuss over the Keronix affair had long since died down and the company had been growing apace. They had followed up on the instantaneous success of their first CPU, the NOVA, by producing a whole line of compatible NOVAs, and while those computers kept selling and selling, they had created another and generally more powerful line of machines, called the Eclipse. Like the NOVA, it was a hit. The Eclipse line was just beginning to grow, however, when the leader of the Eclipse Group began to leave his team. He went away to invent the next kind of Data General computer. As it developed, this became a grand project. Among other things, it would solve one of the important technical questions then on the horizon. This problem lay in the question of how best to enlarge the minicomputer's "logical-address space." This was the problem that superminis like VAX would address, by achieving what one wag called "thirty-two-bit-hood."

Computers, it is often said, manipulate symbols. They don't deal with numbers directly, but with symbols that can represent not only numbers but also words and pictures. Inside the circuits of the digital computer these symbols exist in electrical form, and there are just two basic symbols — a high voltage and a low voltage. Clearly, this is a marvelous kind of symbolism for a machine; the circuits don't have to distinguish between nine different shades of gray but only between black and white, or, in electrical terms, between high and low voltages.

Computer engineers call a single high or low voltage a bit, and it symbolizes one piece of information. One bit can't symbolize much; it has only two possible states, so it can, for instance, be used to stand for only two integers. Put many bits in a row, however, and the number of things that can be represented increases exponentially. By way of analogy, think of telephone numbers. Using only four digits, the phone company could make up enough unique numbers to give one to everybody in a small town. But what if the company wants to give everyone in a large region a unique phone number? By using seven instead of four digits, Ma Bell can generate a vast array of unique numbers, enough so that everyone in the New York metropolitan area or in the state of Montana can have one of his own.

Inside certain crucial parts of a typical modern computer, the bits — the electrical symbols — are handled in packets. Like phone numbers the packets are of a standard size. IBM's machines have traditionally handled information in packages 32 bits long. Data General's NOVA and most minicomputers after it, including the Eclipses, deal with packages only 16 bits long. The distinction is inconsequential in theory, since any computer is hypothetically capable of doing what any other computer may do. But the ease and speed with which different computers can be made to perform the same piece of work vary widely, and in general a machine that handles symbols in chunks of 32 bits runs faster, and for some purposes — usually large ones — it is easier to program than a machine that handles only 16 bits at a time.

In this case, the main issue was the computer's storage system. Here, in packages of symbolic bits, are kept both information for the computer to manipulate and also many of the instructions that tell the computer what to do with that data. The situation resembles that of a region's telephone system; phone are of no use unless they are distinct one from the other, and an item in a computer's storage system is of no use unless it can easily be found. And the general solution resembles the phone company's; each compartment in the computer's storage has its own "phone number," its own unique symbol, known as an address. A 16-bit machine can directly generate symbolic addresses only 16 bits long, which means that it can hand out to storage compartments only about 65,000 unique addresses. A true 32-bit machine, however, can directly address some 4.3 *billion* storage compartments.

Some of Data General's old customers and many potential new ones needed, or soon would, the large "logical-address space" of a 32-bit machine. Although other customers had no such needs as yet, a general feeling held that 32-bithood would become a de facto standard in the industry. You had to produce a 32-bit machine.

It was 1976. By degrees West had taken command of the Eclipse Group. He and his small team were, as he put it, pounding out new 16-bit Eclipses. Meanwhile, the group's former leader and another team of engineers were working on the monumental new machine, which would solve the logical-address problem, among others. The monumental machine had taken on the code name FHP, an abbreviation for "the Fountainhead Project." The team designing FHP repaired to a suite in the Fountainhead Apartments — a local edifice that lends a touch of Miami Beach to the town of Westborough — in order to pursue their work. West's Eclipse Group carried on at headquarters, extending the successful line of Eclipses. And all might have proceeded this way, in relative harmony, but for politics.

Data General had built a new research facility in a place called Research Triangle Park, in North Carolina, a state that had made itself comely to industry, partly by keeping taxes low. While lauding the government of North Carolina, Data General's spokesmen rather bitterly denounced Massachusetts, where taxes of all sorts run high. Edson de Castro himself joined in the criticism. One company spokesman went so far as to lament the fact that Data

General had chosen to grow up in the Bay State. In reporting such comments, none of the Boston papers bothered to point out that the many universities and institutes of technology, existing tax free in Massachusetts, had produced much of the technology and many of the technologists that had made a Data General possible in the first place. But Data General's front office wasn't complaining without reason. In Massachusetts, as elsewhere, 1976 happened to be an election year, and the company was throwing its weight against several propositions on the ballot that threatened to increase both Data General's operating expenses and the personal income taxes of its better-paid employees.

As it turned out, most of the company's favored causes won. No doubt Data General's campaign had an influence. The opening of the facility in Research Triangle Park, North Carolina, and the political campaign, however, had some unfortunate side effects on some of the company's engineers.

For one thing, word had come down that the FHP project was being transferred to North Carolina. Some of the engineers who had been working on that grand new machine refused to pack up their families and go south. At least some of them felt robbed. "You gotta understand," said West later. "FHP was the one thing in the world they wanted to do most, the biggest-thing-the-world's-ever-seen kind of thing. Somebody told those guys that they would have seventy-two uninterrupted hours with the girl of their dreams. The thing they most wanted to do was dangled before them and then pulled away. And some people were pissed."

Then there was the newspaper story. One morning, after FHP's departure had been announced, a number of the engineers at Westborough, who felt of course that they were good, productive, "can-do" engineers, picked up the *Boston Globe* and saw in it an article about themselves. It read in part:

> Speaking to the Boston Security Analysts Society, de Castro said his company, the second largest minicomputer producer in the world, was finding it "a great deal easier to staff [its

research-and-development center] in North Carolina. People
are more willing to move into that area than Boston."

... In addition to this 20 percent cost-of-living difference,
which is a combination of taxes, insurance, housing, food and
other costs, de Castro said the Research Triangle Park area
has "a different feel" than the research facility the company
maintains in Westborough. "The ambition level is different.... There is a can-do attitude and that environment is
contagious."

West remembered the aftermath, laughing low and shaking his
head: "De Castro called us all together and in his inimitable fashion totally confused everyone. He said that the press distorts
things and I hope you know that even if I believed that quote I
wouldn't say it. Then he gave us the march to victory speech and
left." West added, "Morale hit an all-time low at Westborough."

Various accounts of how matters stood among the engineers
there suggest that many believed Westborough's days as an interesting place to work were about over. Sure, said West, people
there could continue to build NOVAs and Eclipses. "But what
fun is that?" Some of those who were going south with FHP
boasted that they were going to the place where the action would
be. And indeed it looked as though the important new machines,
the sorts of projects that a significant number of Westborough's
engineers wanted to work on, would thereafter be pursued in
North Carolina.

Some of the engineers who had chosen New England over FHP
fell under West's command, more or less. And the leader of the
FHP project suggested that those staying behind make a small
machine that would solve the 32-bit, logical-address problem and
would at the same time exhibit a trait called "software compatibility."

Some of those who stayed behind felt determined to build
something elegant. They designed a computer equipped with
something called a mode bit. They planned to build, in essence,
two different machines in one box. One would be a regular old

16-bit Eclipse, but flip the switch, so to speak, and the machine would turn into its alter ego, into a hot rod — a fast, good-looking 32-bit computer. West felt that the designers were out to "kill North Carolina," and there wasn't much question but that he was right, at least in some cases. Those who worked on the design called this new machine EGO. The individual initials respectively stood one step back in the alphabet from the initials FHP, just as in the movie *2001* the name of the computer that goes berserk — HAL — plays against the initials IBM. The name, EGO, also meant what it said.

The people working on EGO, nominally working for West but in fact amenable then to no one's control, truly labored. They worked nights. They worked weekends. They argued hotly with each other. "It was the most incredible, soaring experience of my life," said one of them later. And they worked with astonishing speed. Within two months they had a fairly complete specification. Then they took it to de Castro.

To a disinterested observer it might have seemed obvious that Data General wasn't going to field both EGO and North Carolina's machine. The costs of supporting the two radically new CPUs would be prohibitive. Data General could sensibly afford only one major new machine, and Data General is almost always sensible about money. According to West, and others, de Castro told him and his troops to work out their differences with North Carolina. They didn't. To some of the engineers who worked on EGO, what ensued was a "war," and the first open battle, which was fought at a Howard Johnson's motor inn down south, was "the big shoot-out at HoJo's." Carl Alsing, who was not a participant but an interested observer, said: "I imagine the great EGO wars as a pen-and-ink drawing. Snarling engineers are shown hurling complexities at each other."

Which processor was better? Which deserved to receive the support and resources of the company? Those were the stakes.

Fights like these often take place inside computer companies. So it is said. Here the winner was foreordained. When it came to rating the talents of the engineers, everyone had to agree that some of the company's finest stood behind EGO, but North Carolina's leader was generally acknowledged to be the company's stellar designer after de Castro. Moreover, the company had made a substantial investment in North Carolina as a research-and-development facility, and Data General is a company that likes to see its investments pay off without undue delay. "EGO was a group of five people, FHP had fifty. You're not gonna kill a group of fifty. The company's not gonna send FHP to North Carolina and then do EGO," said West. But that was later on. At this time he argued strongly for EGO. At a meeting before de Castro in September 1977, the two sides traded promises. Westborough would do EGO in a year. Well, then, North Carolina would do a version of FHP in a year. West remembered: "De Castro looked around the room and said, 'It's a dilemma.' It's a famous quote for him, never heard before or since. I said, 'Okay, we'll stop building EGO,' and de Castro walked out of the room."

West told himself that his team hadn't lost; they'd merely retreated from a battle they couldn't win. Thinking of the promises North Carolina had made, West explained later on: "They'd signed up to do the impossible. We weren't signed up to do anything. Right then, that looked like a pretty good position to be in." But EGO's designers took it hard.

Rosemarie Seale, the Eclipse Group's secretary, watched the proceedings from afar. "Engineering is a man's world. I don't know how much territorial interests have to do with it, but they're all fighting for that piece of the pie. There are some who can't admit it, but all of them are," she said. She was sad for her troops. She had made special arrangements to get the document describing EGO nicely typed up. "They wanted it to look really good for de Castro." She had wished them luck when they had gone upstairs with it. Now she watched them trudge back down the hallway. "I knew the minute they hit the department. Such depression. It was terrible, absolutely terrible. Ed de Castro didn't want them to do it."

Certain of the engineers now entered what West called "the first off-the-wall period." A few quit. Others went on vacation immediately. Still others spent the next couple of weeks playing a game called Adventure, in which you travel by computer into an underground world, wandering through strange, awful labyrinths, searching for treasure that's guarded and sometimes snatched away by dragons, dwarfs, trolls and a rapacious pirate who mutters: "Har. Har."

The period after EGO got killed was remembered as West's speechifying period. "Speeches of rising and falling expectations," one engineer called West's orations. West told his group that from now on they would not be engaged in anything like research and development but in work that was 1 percent R and 99 percent D; they'd pound out Eclipses and put money on the bottom line and that was that and if they wanted to build "sexy" machines that "the technology bigots" would like, then they'd have to look elsewhere. Then West turned around and said that although they did not have the charter to build a new 32-bit machine, they could still have some fun and challenge; they'd create a 16-bit minicomputer faster by a factor of two, or maybe even four, than any the world had ever seen. The name of this project was Victor, the Mature Eclipse. "Victor was a canard," said West. But it gave his people something to do.

Dreams of EGO died hard. According to West, some software engineers at Westborough had felt despondent about EGO's demise; they faced the prospect of having no major new CPUs to write system software for. They liked EGO. Thus bolstered, EGO's creators got West to try to attempt an EGO revival. West got the impression from the vice president of engineering that EGO might win de Castro's approval this time. But it didn't. "De Castro basically said, 'Do something to extend the addressing capability of Eclipses, but don't use a mode bit,' " West remembered. West made more speeches of rising and falling expectations. The second off-the-wall period hit their corner of the basement. Privately, West felt very angry.

"No mode bit." It seemed to him that de Castro was asking the Eclipse Group to work with only one hand. West calmed down directly, though. "De Castro's always very vague," he reasoned, "but when he says something, you don't figure he's just shooting off his lip. You find out what's wrong with mode bits." West asked around. He came to the conclusion that, once employed, mode bits tend to proliferate in a company's product line, leading toward unnecessary development costs. "You get hung up in your own underwear." But de Castro had said only what he didn't want. What did he want them to do?

When FHP had gone south there had been some talk of "a 32-bit Eclipse." Some time before that, a person in Marketing had asked some members of the Eclipse Group if they couldn't just tweak a few things inside an Eclipse and give it the extended logical-address space of a 32-bit machine. Back then, they had brushed off the suggestion; it wasn't that simple. Now West virtually camped out in this executive's office. He wanted to know what it was that customers most desired. VAX had by now gone to market, with all the usual hoopla, and was selling briskly. Other minicomputer companies had entered 32-bit machines in the supermini sweepstakes. And a lot of Data General's old customers wanted one, but it seemed clear that they would also want a machine that displayed the quality called software compatibility.

An old-fashioned automobile that can be started only with a crank requires a person to make a series of adjustments more or less directly to the engine; in a modern car, of course, you need only turn a key and a system of electrical and mechanical devices does the rest. In the modern computer, software has developed in such a way as to fill this role of go-between. On one end you have the so-called end user who wants to be able to order up a piece of long division, say, simply by supplying two numbers to the machine and ordering it to divide them. At the other end stands the actual computer, which for all its complexity is something of a brute. It can perform only several hundred basic operations, and long division may not be one of them. The machine may have to be instructed to perform a sequence of several of its basic operations in order to accomplish a piece of long division. Software — a series of what are known as programs — translates the end user's wish into specific, functional commands for the machine.

There are two general kinds of computer programs. The users write one kind themselves or hire consultants to write them. These "user programs" may look as though they are telling the computer what to do step by step, and a relatively simple program — for calculating a company's payroll, for instance — may look forbiddingly long and complex. But this is nothing like the length and complexity that the program would have if it really did work directly on the machine. Instead, nowadays, a series of other programs that are stored inside the computer breaks down commands such as "divide" into several more basic commands that the machine is equipped to obey. These intermediary programs, which serve as translators of user programs, are known collectively as "system software." It is usually the manufacturer's responsibility to create system software, and the customer's to buy it when purchasing a new machine.

By the mid-1960s, a trend that would become increasingly pronounced was already apparent: while the expense of building a computer's hardware was steadily declining, the cost of creating both user and system software was rising. In an extremely bold stroke, IBM took advantage of the trend. They announced, in the mid-sixties, all at one time, an entire family of new computers — the famous 360 line. In the commerce of computers, no single event has had wider significance, except for the invention of the transistor. Part of the 360's importance lay in the fact that all the machines in the family were software compatible.

It cost IBM a true fortune and no end of trouble and anxiety to create system software for the 360 line. But all the machines in the family used that same software. So IBM had to create the stuff only once, and thus was able to amortize the cost of its development over all of the many thousands of 360's that it sold. Moreover, any user program that worked on one machine in the family worked on all of them. Users become attached to their programs and system software. Software is expensive. Getting it to function properly often takes time. Software that works is precious. Users don't idly discard it. Obviously, this reluctance can present problems for computer manufacturers: how do you get customers to buy bigger, better machines? Total software compatibility made it easy for customers to do what IBM wanted them to do, which was to buy several different kinds of 360 computers. A customer could buy a small one now and later on buy a bigger one, or vice versa, without having to re-create any software. Software compatibility strengthened IBM's already tight grip on its customers: they weren't likely to forsake IBM and take their business elsewhere when that meant assuming new expenses and problems with software.

Soon every manufacturer of computers was employing some variation of IBM's 360 strategy of software compatibility. Data General had made all of its NOVAs compatible one with the other, and likewise all the Eclipses. Moreover, the designers had made the Eclipse "upwardly compatible" with NOVAs. This meant that while new programs written for Eclipse computers would not run on NOVAs, old programs written for NOVAs *would* work on Eclipses. This sort of compatibility was a useful tool for marketing, for it allowed customers to switch from NOVAs to Eclipses with relative ease — they could do so without discarding all their old software.

Software compatibility is a marvelous thing. That was the essential lesson West took away from his long talks with his friend in Marketing. You didn't want to make a machine that wasn't compatible, not if you could avoid it. Old customers would feel that since they'd need to buy and create all new software anyway, they might as well look at what other companies had to offer; they'd be likely to undertake the dreaded "market survey." And an incompatible machine would not make it easy for new customers to buy both 16-bit Eclipses and the new machine. This was the gist of what West learned. He became increasingly interested.

DEC's VAX was only "culturally compatible" with the line of machines that preceded it. Data General should build a 32-bit machine that was *fully* compatible with Eclipses. "From a marketing point of view what a win that would be!"

Matters proceeded so informally and with such speed that there would never be any way of telling afterward where all of the first technical solutions to the problem of making a 32-bit Eclipse came from. But it was West who gathered them together. Soon he began an in-house PR campaign that would not end for a long time.

Carl Alsing participated in the beginnings of West's proselytizing. Among other acts, Alsing gave the new, unbuilt machine the code name Eagle. Mostly, though, he observed. Alsing was by temperament a watcher, a moviegoer. He had been with West longer than anyone else in the Eclipse Group. He felt that he knew West and then again he felt that he didn't. West launching Eagle was to Alsing something worth watching. For example, the meeting West set up with the vice presidents of engineering and software. West took Alsing along. The way it looked to Alsing, West brought two proposals to the VPs. One of these was an obvious loser and the other was Eagle. "West's letting them pick Eagle," thought Alsing, and he smiled.

"West's never unprepared in any kind of meeting. He doesn't talk fast or raise his voice. He conveys — it's not enthusiasm exactly, it's the intensity of someone who's weathering a storm and showing us the way out. He's saying, 'Look, we gotta move this way.' Then once he gets the VPs to say it sounds good, Tom goes to some of the software people and some of his own people. 'The bosses are signed up for this,' he tells them. 'Can I get you signed up to do your part?' He goes around and hits people one at a time, gets 'em enthused. They say, 'Ahhh, it sounds like you're just gonna put a bag on the side of the Eclipse,' and Tom'll give 'em his little grin and say, 'It's more than that, we're really gonna build this fucker and it's gonna be fast as greased lightning.' He tells them, 'We're gonna do it by April.' That's less than a year away, but never mind. Tom's message is: 'Are you guys gonna do it or sit on your ass and complain?' It's a challenge he throws at them. So he basically made us stop moaning about the demise of Westborough."

Alsing went on: "West brought us out of our depression into the honesty of pure work. He put new life into a lot of people's jobs, I think."

Not everyone associated with the Eclipse Group liked the looks of this proposed new machine. They thought it would be just a refinement of the Eclipse, which was itself a refinement of the NOVA. "A wart on a wart on a wart," one engineer said. "A bag on the side of the Eclipse." Some even said that it would be a "kludge," and this was the unkindest cut. *Kludge* is perhaps the most disdainful term in the computer engineer's vocabulary: it conjures up visions of a machine with wires hanging out of it, of things fastened together with adhesive tape.

So some engineers dropped out of the project right away. To the remainder and to new recruits, West preached whatever gospel seemed most likely to stir up enthusiasm. It would be an opportunity for them to "get a machine out the door with their names on it." When Alsing came up with the code name, Eagle, West felt pleased, because, he said, it was impossible to say "Eagle" without it sounding as if you were saying "EGO." Not that West cared himself; he did not feel at all vengeful toward North Carolina. But some others did. There would also be something here for technology bigots. Eagle would not be, as the saying goes, "a clean sheet of paper"; but there would be some clean corners to work in. It might look at first glance like a VW Beetle, but think what they could put under the hood. This wasn't going to be just a slight variation on the Eclipse CPU, but a wholly new and fast machine that would happen to be compatible with Eclipses. It would put money on the bottom line, of course. Lots of money. They were going to do it in record time, because the company needed this machine desperately. And when they succeeded, they would be heroes.

If, after EGO had been canceled, you had left the Eclipse Group moping in their corner of the basement, and had returned a year later, you surely would have failed to recognize the place. At some moments, their corner of the building now had the air of a commuter train about it, and at others, the silent intensity of a university library on the eve of exams — many new young faces peering into cathode-ray tubes, leafing through fat documents. In the conversation around there you heard words and phrases such as these: A *canard* was anything false, usually a wrongheaded notion entertained by some other group or company; things could be done in ways that created *no muss, no fuss,* that were *quick and dirty,* that were *clean. Fundamentals* were the source of all right thinking, and weighty sentences often began with the adverb *fundamentally,* while *realistically* prefaced many flights of fancy. There was talk of *wars, shootouts, hired guns* and people who *shot from the hip.* The *win* was the object of all this sport and *the big win* was something that could be achieved by *maximizing* the smaller one. From the vocabulary alone, you could have guessed that West had been there, and that these engineers were up to something.

Rosemarie Seale, the group's main secretary, felt excitement in the air. She, for one, took West's exhortations to heart. She resolved to do whatever she could in order to keep bureaucratic and trivial affairs from distracting these youngsters on their crucial mission. Her spirit never flagged, but from time to time she did wonder why, if this project was so important to the company, so few people in other departments seemed to recognize the fact. Why, for instance, was it allowed that the mailroom be moved in the midst of the project, creating the risk that vital mail would be held up? To prevent that sort of small catastrophe, for several weeks she would go to the mailroom daily and sort through the mail herself. Likewise, why would the carpenters be allowed to come in during a particularly delicate phase of the Eagle project and totally remodel the Eclipse Group's office space?

The answer, one possible answer, was that West had two ways

of describing Eagle. One way made it sound important and glorious; the other, like something routine. West explained: "You gotta distinguish between the internal promotion to the actual workers and the promoting we did externally to other parts of the company. Outside the group I tried to low-key the thing. I tried to dull the impression that this was a competing product with North Carolina. I tried to sell it externally as not much of a threat. I was selling insurance; this would be there if something went wrong in North Carolina. It was just gonna be a fast, Eclipse-like machine. This was the only way it was gonna live. We had to get the resources quietly, without creating a big brouhaha, and it's difficult to get a lot of external cooperation under those circumstances."

West went on, theorizing now: "The company would have been just as happy if we hadn't done it. De Castro thought he had it covered in North Carolina. But once you got somebody saying, 'Hey, we want to do this,' then you're in a position where you gotta say no, and that's a different proposition. A bunch of good engineers were getting ready to quit from being told no too many times already; that's another problem. I went to de Castro about Eagle at some point. I said, 'We'll do it in a year,' and he may have said, 'Okay.' But it was clear we had to do it in a year to have any chance."

Years before, according to local legend, West's mentor, the former leader of the Eclipse Group, had said that he could build a NOVA on a single printed-circuit board. Told that he was suffering from delusions of grandeur, he did the job at home on his kitchen table and wound up producing the best-selling of all the NOVAs. Before that, de Castro and two of the other founders had cut loose from their old company, in order to build a new computer. West was doing nothing so extreme. He seemed to have the full support of Data General's vice president of engineering, Carl Carman. He was given money to hire new recruits. He had de Castro's laconic permission to give Eagle a try. But even after the project had begun, West would say (for example), "De Castro won't take a piece of paper from me on this machine." He would remark, "There's a lot of people pretending that this project doesn't exist." Some others in the group expressed the same contradictory feeling: that they were building a machine absolutely essential to the company but were doing it all on their own. "I think we're doing it in spite of Data General," one of the older hands in the group said. The circumstances fostered such an attitude. So did West. Deliberately, somewhat surreptitiously, he separated his team from the rest of the company.

"We're building what I thought we could get away with," West said.

3

BUILDING A TEAM

THE BASEMENT of Westborough, subterranean at the front of Building 14A/B and at ground level in back, was one of the places in Data General's widening empire where machines were conceived, designed, labored over in prototypes, and sometimes brought to life. Tom West led the way down into this region one night, through confusing corridors. I would try to learn my way around by noting small landmarks — a copying machine at a corner, a bulletin board trimmed with company news about dental programs and new Data General disk drives. Off some of the hallways were mysterious doors, locked up and bearing signs that said in loud letters RESTRICTED AREA.

Then the hallways opened and all around under fluorescent light lay fields of cubicles without doors. Their walls — made of steel, some of them covered with cream-colored cloth — did not reach the ceiling, but stood about five and a half feet high. You could look over them. They created no privacy. Most of the cubicles were empty now, but each would contain one person during the day. Most had a desk with a computer terminal on it, and a little bookcase. Some held a drafting table and many had a houseplant or two. Many green plants poked their heads, like periscopes, above the tops of the cubicles' walls. "The great statement," said West, gesturing at the foliage, wearing a little grin that puckered one cheek.

The arrangements looked temporary, and in fact they were. As one of the company's PR men explained, cubicles laid out as in a maze allow a greater density of workers per square foot than real offices with doors. Easily movable walls allow management to tinker with that area-to-people ratio without incurring enormous expense — for instance, to create fully enclosed offices where cubicles have been when it turns out that some jobs are most efficiently performed behind doors. It was said that the company's vice president for manufacturing could turn Westborough into a factory overnight, and maybe the joke had some substance. The last headquarters — compared to which this one was plush — had in fact been turned into a factory.

Westborough seemed designed for quick changes. Smiling his wry smile, West offered some additional theories: "We can change it all around. It keeps up the basic level of insecurity. . . . It's basically a cattle yard. . . . What goes on here is not part of the real world."

"How so?"

"Mmmmmmmmmmh. The language is different."

Some of it was, and a phrase book, such as the Penguin *Dictionary of Computers,* could be useful. *ECO* — each letter pronounced — meant "engineering change order." Hence this remark: "A friend of mine told his girlfriend they had to ECO their relationship." *Give me a core dump* meant "Tell me your thoughts," for in the past, when computers used "core memories," engineers sometimes "dumped" the contents of malfunctioning machines' storage compartments to see what was wrong. A *stack* is a special small compartment of memory, a sort of in-box inside a computer; it holds information in the order in which the information is deposited and when it gets overfull, it is said to "overflow." Hence the occasional complaint, "I've got a stack overflow." "His mind is only one stack deep," says an engineer, describing the failings of a colleague, but the syntax is wrong and he rephrases, saying: "See. He can push, but when it comes time to pop, he goes off in all directions" — which means that the poor fellow can receive and understand information but he can't retrieve it in an orderly fashion.

The basement, it seemed, was never empty. Even in the small hours of the morning someone would be sitting in a little pool of light in a cubicle, working. By day, the place held a throng. I saw them all collected once, out back in the parking lot, during a fire drill. I counted only a couple of black faces, but I saw many women, many of them in skirts, and I presumed most were secretaries, because I knew that here as in the industry generally female engineers were scarce. Men were numerous. Most looked as though they were in their twenties. Only a few wore jackets and ties; the rest were dressed casually and, on the whole, neatly. Once, down in the basement, I saw an engineer who wore not just long but shaggy hair and who was wearing Army-surplus clothes. He was slouching off down a corridor carrying a canteen cup. His appearance was sufficiently unusual that one of West's team took pains to point him out to me.

West led the way into the Eclipse Group's quarters. They were indistinguishable from any other group's, except at night perhaps, during the Eagle project, when, as a rule, more lamps burned on in their cubicles than elsewhere in the basement. You could tell that an engineer enjoyed some rank if he had an office with a door. West had one of these. It was tiny and windowless. A thick, jacketed pipe and a steel girder descended through it, down the face of a cinder-block wall. There were some gray metal chairs, a gray metal bookcase, a couple of small gray metal tables and a gray metal desk, the top of which was absolutely clean save for a single stack of papers with their edges perfectly squared. A Magic Marker board, at the moment displaying some incomprehensible diagram, hung on one wall. For adornments there were an old clock in a beautiful oak case, and on the wall behind West's back, a picture of a square-rigged sailing ship. On the wall beside him hung several photographs of computers.

Physicians hang diplomas in their waiting rooms. Some fishermen mount their biggest catch. Downstairs in Westborough, it was pictures of computers.

When engineers finished a project and word of approval came down from the offices of the executives and the various arms of the company got ready to announce the machine to the world that spends so lavishly on computing equipment, then the Marketing Department would usually hand out to each person who had helped to create the new computer a framed photograph of it. Several of these sorts of pictures hung in West's office. West had a saying: "The game around here is getting a machine out the door with your name on it." One of the photographs on his wall showed a model of Data General's first Eclipse computer, and, apropos of his remark, printed on the machine was a list of eight names. West's was among them and so was Carl Alsing's. An identical photograph hung in Alsing's cubicle.

The same image, of that first Eclipse, also sat on a windowsill in an office at another computer company. It belonged to an engineer who had worked for Data General back in the days when the Eclipse was being built. The picture was just a photo of an immobile plastic box, but the former Data General engineer gazed at it, and he smiled. "It was a lot of fun, a lot of pressure. With the Eclipse there was a tremendous amount of team spirit. We were going twenty-four hours a day debugging that prototype, breathing on it hard to make it come to life.

"West did do an awful lot of the debugging. I would say he's an excellent engineer. I really think that Tom was very much of a problem solver. It was decided that the Eclipse should have error-correction code. What was that? There wasn't that much written about it at that time. Tom went and learned about it and came up with how to do it. Some reports I hear about Tom — I hear that he's very different in a management role. I hear in his management role he's a very tough guy, very closed, but he was an easygoing guy then, though hardworking.

"Tom used to have these all-day picnics at his house. He'd roast a pig, there'd be a keg of beer. He was a hell of a nice guy! I enjoyed him a lot."

Many people recalled West's annual pig roasts. Enough people attended, said another old friend, that the odds lay in favor of a child being born during the fete, and the guest list included painters and writers, musicians, footloose young folks and also computer people. West is remembered, his face all lit up, moving among the throng, surveying his wide and disparate acquaintances. "He was so happy and funny and warmhearted," said another regular at the roasts.

West came to Data General in 1974, joining Carl Alsing and the other engineers who were attempting to bring the first Eclipse to life. To Alsing, West appeared to be just a good, competent circuit designer, but strikingly adept at finding and fixing the flaws in a computer. "A great debugger," Alsing considered him. "He was so fast in the lab I felt I was barely adequate to hold the probes of the oscilloscope for him." Alsing took a shine to his new colleague almost at once. The morning after an Eclipse Group party, he and West, on the spur of the moment, took a trip to Provincetown on Cape Cod. Alsing was amazed at how easily West found his way among strangers and how he seemed to be able to identify the most interesting bars by just glancing through their doorways. West possessed that town for an evening, as if he'd lived there all his life and were showing Alsing around.

One time Alsing stayed up all night in a lab programming a batch of ROMs, or "read-only memory" chips. West found Alsing still at this work the next morning, and laughing, he cried out: "Alsing! You're a ROM-driven man!" Then West made up a song by that title to the tune of "John Henry." West was always inventing catchy expressions, and "ROM-driven" was Alsing's favorite. The contents of a ROM, once programmed into the chip, cannot be altered or erased; the information can only be "read." *ROM-driven:* it opens up the ancient question of predestination and free will. Later, Alsing would wonder how the phrase applied to his friend. So would West himself, apparently, for one day downstairs, during the Eagle project, he would ask — laughing his nervous-sounding laugh, not the hearty one of the pig roasts — "Doesn't anything around here happen by accident?"

From West, Alsing was able to glean only a few biographical facts: that he had gone to Amherst College, majoring in physics there; that he had worked for the Smithsonian Institution afterward, building digital clocks, among other things, and traveling a lot; that he had quit that job just like that, after seven years, and had virtually taught himself computer engineering while working for RCA. There was a little more: West's father was an important man, one of the most senior executives of AT&T; West had a wife and daughters; West played the guitar well, and personally knew many famous folksingers. Alsing listened to his stories. West told him that one night, while he was traveling through Mozambique on business for the Smithsonian, he got out of a Landrover and yelled into the darkness: "Massachusetts! Massachusetts!"

"I thought someone might hear me," West explained, "and someday there'd be a bunch of kids running around out there named Massachusetts."

Alsing clapped his hands and laughed and laughed. He always wanted to hear more.

West had been to places that Alsing never dreamed he himself would visit. Alsing could not help envying him a little, not least of all because West seemed so free. West was, to Alsing then, like that mysterious stranger just passing through town. West told him about quitting his job with the Smithsonian on the spur of the moment, and of a troupe of latter-day gypsies — a band of youngsters on the roam who had camped in a field near his house. Alsing was left feeling that if those gypsies passed through town again, West might go away with them. When West talked about his music, Alsing got the same feeling; he fully expected to come to work some morning and find that West had departed for good, most likely without leaving a forwarding address, and the idea charmed Alsing — and also made him feel a little sad, of course.

But it didn't turn out that way. When the first Eclipse went to market and the group's original leader began to take leave, in order to work on FHP, West asked that he be given command of the team. West seemed to Alsing the logical choice for the job: "He was the smartest guy around." But Alsing was very surprised that West wanted the position.

As Alsing remembered, West was told in effect that it was out of the question. West had been asked weeks before to design a piece of equipment called an IOP, and he had not done a lick of work on it yet. What made him think he could run the Eclipse Group?

West went into his office then and closed his door for about seven weeks.

West and Alsing usually went out for coffee in the middle of the morning, but not now.

Alsing poked his head into West's office. "Coffee, Tom?"

"Go away, Carl," West replied.

Alsing tried on another day. Without looking up from his work, West said, in a flat, calm voice, "Get out, Alsing."

Alsing sensed that there was nothing personal in these rebuffs, and he found it impossible to get angry at West. After seven weeks, West emerged, the completed design for the IOP in his hands. Little by little after that, he assumed command of the Eclipse Group. Reflecting some years later on those seven weeks of West's labor, Alsing said: "The day Tom went into his office to do the IOP was the day he started getting tough. I think it was the day when he started to care."

Over the next several years, successive generations of engineers joining the group would know less of West than their predecessors had, until finally, by the time of Eagle, new recruits would know almost nothing about him at all. They'd know nothing of pig roasts; that custom had lapsed. Their view of West would be restricted mainly to chance encounters in hallways. West would come down a corridor dragging the knuckles of one hand along the wall, and often he would pass right by members of his own team, without, it seemed, even noticing them. For their part, most gave up trying to greet him. The distant, angry look on his face warned against it.

Occasionally, some of the Eclipse Group wondered about their leader.

"We've heard that he worked for the CIA."

"Wasn't he a folksinger?"

"A lot of guys think he's a speed freak."

"West," said one young engineer, "is a prince of darkness."

He had changed. There was no denying that. Though remaining as close to him as anyone in the basement, Alsing rarely saw him after work anymore. Almost no one from Westborough did, it seemed. West was Alsing's boss now; that was part of it. But he wasn't as obviously happy and merry as he had been, and his jokes, though Alsing still found them witty, tended now toward the sardonic. He'd smile with one side of his mouth, whereas he used to grin with all of it. Every so often, Alsing did see flashes of the old West. This was true especially in West's campaign after EGO, his assiduous canvassing for Eagle, his successful attempt to rouse various parts of the basement from depression. To be sure, most of the humor and merriment was still missing, and this fellow of formerly eclectic passions now seemed completely single-minded; but West's campaign reminded Alsing of their adventure years before in Provincetown. West rarely lent his enthusiasm to "someone else's trip," Alsing noted. "But if it's his own or he makes it his own . . ."

West still had a way of making ordinary things seem special; in this case a 32-bit Eclipse was being transformed into the occasion for an adventure. West's ardor for it seemed to spread the way his neologies did. Others besides Alsing felt infected.

For Rosemarie Seale, the main excitement began after EGO was canceled and everyone but West seemed ready to pack it in. "Tom's obviously made some decision which I know nothing about," she said. "He's decided he isn't gonna take his bat and ball and go home." Later, she would say: "I wanted to work for him. I could have gotten more pay elsewhere. I didn't understand it all, but I knew I wanted to work for him. I wanted to be part of that effort."

Rosemarie is a petite, brown-haired woman in her middle years. She speaks rapidly and punctuates most statements with a quick, low laugh which makes it seem as though she's chuckling while she talks. "I grew up in a poor family, in Depression times. I went to secretarial school in Boston. I brought up a family. I got divorced. I traveled a bit. I was a stupid, ignorant girl when I was young, and I think I've learned a few things, but probably not many." In 1976 Rosemarie was working for an insurance company, supervising the underwriting files, a job that held her interest for exactly one month. The files had been a mess, but once she'd gotten them cleaned up, the work became completely routine. Then she saw a help-wanted ad for Data General in a newspaper. It asked:

ARE YOU BORED?

"It was speaking to me!"

She was assigned to the Eclipse Group, which was tiny then and had never had a secretary. The engineers "found" her a desk, as she put it. She opened the team's lone filing cabinet and found nothing in it, except for a couple of rolls of toilet paper. No list of the group's members existed. She went from one engineer to another, asking, "Do you have *any* idea who you work for?" It was the beginning of a long romance.

To Rosemarie, the Eagle project was like a gift. She had so much to do every day: budgets to prepare, battles to fight with one department or another, mail to sort when the mailroom was untimely moved, phones to answer, documents to prepare, paychecks to find and deliver on time, the newcomers to attend to ("Would they have a place to sit — and the conditions weren't the best, you know — and would they have a pencil?"). Each day brought another small administrative crisis. "I was doing something important," she said.

Rosemarie did not always realize that she was having a good time, of course. Having trained all the team's newcomers to tie their shoes, in an administrative sense, she began to feel that they regarded her as a surrogate mother, which annoyed her, until she decided: "You are what you are. I might as well enjoy that, too." She always hated answering the phones, though. Once in a while she even talked about quitting. Sympathizing, Alsing once asked her why she didn't just do it.

"I can't leave," she answered. She flashed a furtive grin at Alsing, and nodding her head toward West's office door, she said, in a lowered voice: "It's like one of those terrible movies. I just have to see how it comes out. I just have to see what Tom's gonna do next."

So there was another watcher. Alsing was delighted.

Convinced that Eagle would be a wart, a bag, a kludge — and suspicious that it would go the way of EGO and Victor — some of the brightest hardware engineers around expressed no interest in joining the project. Others went along, some reluctantly at first, and by the very early spring of 1978 West had gathered the makings of a team. He had Rosemarie and Alsing and about a dozen other experienced engineers, who had worked for him before. For a time West thought that their numbers would suffice, but really they were just a cadre. It became obvious, when they started designing the "logic" of the new machine, that such a tiny group would never be able to produce a computer like this in a year. "We need more bodies," said West to Alsing, and Alsing agreed.

North Carolina's leaders had assembled a large crew mainly by luring experienced engineers away from Westborough and other companies. But around this time videotape was circulating in the basement, and it suggested another approach. In the movie, an engineer named Seymour Cray described how his little company, located in Chippewa Falls, Wisconsin, had come to build what are generally acknowledged to be the fastest computers in the world, the quintessential number-crunchers. Cray was a legend in computers, and in the movie Cray said that he liked to hire inexperienced engineers right out of school, because they do not usually know what's supposed to be impossible. West liked that idea. He also realized, of course, that new graduates command smaller salaries than experienced engineers. Moreover, using novices might be another way in which to disguise his team's real intentions. Who would believe that a bunch of completely inexperienced engineers could produce a major CPU to rival North Carolina's?

"Shall we hire kids, Alsing?" said West.

For a couple of weeks he and Alsing discussed the idea. To make it work they'd have to hire the very best new engineers they could find, ones who would know more about the state of the art in computers than they did. They told each other that they'd have to be sure not to turn away candidates just because the youngsters made them feel old and obsolete; on the contrary, those were the candidates they'd have to welcome. Smiling, West allowed that if they did this, they might be hiring their own replacements — their own assassins. Even if they did hire prodigies, of course, the scheme might not work. Maybe you couldn't build a major CPU with kids. It was awfully risky. It was a compelling idea.

Between the summer of 1978 and the fall of that year, West's team roughly doubled in size. To the dozen or so old hands — old in a relative sense — were added about a dozen neophytes, fresh from graduate schools of electrical engineering and computer science. These newcomers were known as "the kids." West was the boss, and he had a sort of adjutant — an architect of the electronic school — and two main lieutenants, each of whom had a sublieutenant or two. One lieutenant managed the crew that worked on the hardware, the machine's actual circuitry, and the members of this crew were called, and called themselves, "the Hardy Boys." The other main part of the team worked on microcode, a synaptic language that would fuse the physical machine with the programs that would tell it what to do. To join this part of the group, which Alsing ran, was to become one of "the Microkids." There were also a draftsman and some technicians. The group's numbers changed from time to time, generally diminishing, as people dropped out. But usually they totaled about thirty.

How was it to be one of "the kids"? You were in no danger of being fired, but you didn't know that, and besides, when you are brand-new in a job you want to make a good impression right from the start. So you set out to get to know your boss, as Hardy Boy Dave Epstein did. You walk into his office and say, "Hi, I'm Dave," and you begin to extend your hand. Epstein would never forget that experience: "West just sat there and stared at me. After a few seconds, I decided I'd better get out of there."

Going to work for the Eclipse Group could be a rough way to start out in your profession. You set out for your first real job with all the loneliness and fear that attend new beginnings, drive east from Purdue or Northwestern or Wisconsin, up from Missouri or west from MIT, and before you've learned to find your way to work without a road map, you're sitting in a tiny cubicle or, even worse, in an office like the one dubbed the Micropit, along with three other new recruits, your knees practically touching theirs; and though lacking all privacy and quiet, though it's a job you've never really done before, you are told that you have almost no time at all in which to master a virtual encyclopedia of technical detail and to start producing crucial pieces of a crucial new machine. And you want to make a good impression. So you don't have any time to meet women, to help your wife buy furniture for your apartment, or to explore the unfamiliar countryside. You work. You're told, "Don't even mention the name Eagle outside the group." "Don't talk outside the group," you're told. You're working at a place that looks like something psychologists build for testing the fortitude of small animals, and your boss won't even say hello to you

New and old hands told the same story. Chuck Holland: "I can hardly say I do anything else now. It takes about three days to get Eagle out of my mind, so if you have a three-day weekend, you're just sorry to see Monday come." Microkid Betty Shanahan, the group's lone female engineer: "You can end up staying all night. You can forget to go home and eat dinner. My husband complained that the last three times he's had to do the laundry." Jon Blau: "I've had difficulty forming sentences lately. In the middle of a story my mind'll go blank. Pieces of your life get dribbled away. I'm growing up, having all those experiences, and I don't want to shut them out for the sake of Data General or this big project." Jim Guyer, a Hardy Boy and an old hand at age twenty-six, said: "I like my job, it's great, I enjoy it. But it's not what I do for recreation. Outside of work, I do other things, like rock-climbing and hiking." Guyer paused. A thought had just occurred to him. "I haven't done any of that lately. Because I've been working too much."

But where did the relish in their voices come from?

At the start of the project, a newcomer could expect to earn something like $20,000 a year, while a veteran such as Alsing might make a little more than $30,000 — and those figures grew enormously at Data General and elsewhere over the next few years. But they received no extra pay for working overtime. The old hands had also received some stock options, but most seemed to view the prospect of stock as a mere sweetener, and most agreed with Ken Holberger, sublieutenant of Hardy Boys, who declared, "I don't work for money."

Some of the recruits said they liked the atmosphere. Microkid Dave Keating, for instance, had looked at other companies, where de facto dress codes were in force. He liked the "casual" look of the basement of Westborough. "The jeans and so on." Several talked about their "flexible hours." "No one keeps track of the hours we work," said Ken Holberger. He grinned. "That's not altruism on Data General's part. If anybody kept track, they'd have to pay us a hell of a lot more than they do." Yet it is a fact, not entirely lost on management consultants, that some people would rather work twelve hours a day of their own choosing than eight that are prescribed. Provided, of course, that the work is interesting. That was the main thing.

A couple of the Microkids were chatting. They talked about the jobs they had turned down.

"At IBM we wouldn't have gotten on a project this good. They don't hand out projects like this to rookies."

"They don't hand out projects like this to rookies anywhere but Data General."

"I got an offer at IBM to work on a memory chip, to see what could be done about improving its performance. Here I got an offer to work on a major new machine, which was gonna be the backbone of company sales. I'd get to do *computer* design. It wasn't hard to make that choice."

Bob Beauchamp, another Microkid, had come from Missouri. He wore a small red beard. He was perhaps the most easygoing of the recruits. He had wider experience of the world than most, having taken a year off from school to play in a traveling rock band. Beauchamp seemed to be one of those fortunate souls: likable, modest, good-looking and smart. He had compiled an unblemished, straight-A average in graduate school. "I tended to enjoy takin' tests through school. I kinda like to measure myself," he said. "I'd spent five years in college learning, never really doing anything. When I got to Data General, I figured it was time to do something; and also I was new to this part of the world, and by myself, and even on weekends I didn't have a whole lot better to do. I might as well pass the time at work." But Beauchamp had opted to work on a part of the project that turned out to have a low priority. "There was no pressure. I felt out of the mainstream of things. There was intensity in the air. I kinda liked the fervor and I wanted to be part of it." Eventually, a suggestion came down that Beauchamp go to work on some of the machine's microcode. He was essentially offered the chance for some grueling work, and he accepted with alacrity. "I jumped on it, really," he said.

Talk about Tom Sawyer's fence.

There was, it appeared, a mysterious rite of initiation through which, in one way or another, almost every member of the team passed. The term that the old hands used for this rite — West invented the term, not the practice — was "signing up." By signing up for the project you agreed to do whatever was necessary for success. You agreed to forsake, if necessary, family, hobbies, and friends — if you had any of these left (and you might not if you had signed up too many times before). From a manager's point of view, the practical virtues of the ritual were manifold. Labor was no longer coerced. Labor volunteered. When you signed up you in effect declared, "I want to do this job and I'll give it my heart and soul." It cut another way. The vice president of engineering, Carl Carman, who knew the term, said much later on: "Sometimes I worry that I pushed too hard. I tried not to push any harder than I would on myself. That's why, by the way, you have to go through the sign-up. To be sure you're not conning anybody."

The rite was not accomplished with formal declarations, as a rule. Among the old hands, a statement such as "Yeah, I'll do that" could constitute the act of signing up, and often it was done tacitly — as when, without being ordered to do so, Alsing took on the role of chief recruiter.

The old hands knew the game and what they were getting into. The new recruits, however, presented some problems in this regard.

The demand for young computer engineers far exceeded the supply. The competition for them was fierce. What enticements could the Eclipse Group offer to the ones they wanted that companies such as IBM could not? Clearly, West and Alsing agreed, their strongest pitch would be the project itself. Alsing reasoned as follows: "Engineering school prepares you for big projects, and a lot of guys wind up as transformer designers. It's a terrible letdown, I think. They end up with some rote engineering job with some thoroughly known technology that's repetitive, where all you have to do is look up the answers in books." By contrast, Alsing knew, it was thought to be a fine thing in the fraternity of hardware engineers — in the local idiom, it was "the sexy job" — to be a builder of new computers, and the demand for opportunities to be a maker of new computers also exceeded the supply. West put it this way: "We had the best high-energy story to tell a college graduate. They'd all heard about VAX. Well, we were gonna build a thirty-two-bit machine less expensive, faster and so on. You can sign a guy up to that any day of the year. And we got the best there was."

But the new recruits were going to be asked to work at a feverish pace almost at once. They'd have no time to learn the true meaning of signing up on their own. They had to be carefully selected and they had to be warned. Common decency and the fear of having to feel lingering guilt demanded that this be done.

The Eclipse Group solicited applications. One candidate listed "family life" as his main avocation. Alsing and another of West's lieutenants felt wary when they saw this. Not that they wanted to exclude family men, being such themselves. But Alsing wondered: "He seems to be saying he doesn't want to sign up." The other lieutenant pondered the application. "I don't think he'd be happy here," he said to himself. The applicant's grades were nothing special, and they turned him away.

Grades mattered in this first winnowing of applications — not only as an indication of ability but also as a basis for guessing about a recruit's capacity for long, hard work — and with a few exceptions they turned down those whose grades were merely good.

Alsing hoped to recruit some female engineers, but in 1978 they were still quite scarce. Only a few young women applied, and Alsing hired one, who had fine credentials.

When they liked the looks of an application, they invited the young man — it was usually a young man — to Westborough, and the elders would interview him, one by one. If he was a potential Microkid, the recruit's interview with Alsing was often the crucial one. And a successful interview with Alsing constituted a signing up.

Alsing would ask the young engineer, "What do you want to do?"

Exactly what the candidate said — whether he was interested in one aspect of computers or another — didn't matter. Indeed, Alsing didn't care if a recruit showed no special fondness for computers; and the fact that an engineer had one of his own and liked to play with it did not argue for him.

If the recruit seemed to say in reply, "Well, I'm just out of grad school and I'm looking at a lot of possibilities and I'm not sure what field I want to get into yet," then Alsing would usually find a polite way to abbreviate the interview. But if the recruit said, for instance, "I'm really interested in computer design," then Alsing would prod. The ideal interview would proceed in this fashion:

"What interests you about that?"

"I want to build one," says the recruit.

("That's what I want to hear," thinks Alsing. "Now I want to find out if he means it.")

"What makes you think you can build a major computer?" asks Alsing.

"Hey," says the recruit, "no offense, but I've used some of the machines you guys have built. I think I can do a better job."

("West and I have a story that we tell about Eagle machine. But I want to hear this guy tell me part of that story first. If he does, if there's some fire in his eyes — I say 'in his eyes,' because I don't know where it is; if it's there, it's there — but if he's a little cocky and I think we probably want this person, then I tell him *our* story.")

"Well," says Alsing, "we're building this machine that's way out in front in technology. We're gonna design all new hardware and tools." ("I'm trying to give him a sense of 'Hey, you've finally found in a big company a place where people are really doing the next thing.'") "Do you like the sound of that?" asks Alsing.

"Oh, yeah," says the recruit.

("Now I tell him the bad news.")

"It's gonna be tough," says Alsing. "If we hired you, you'd be working with a bunch of cynics and egotists and it'd be hard to keep up with them."

"That doesn't scare me," says the recruit.

"There's a lot of fast people in this group," Alsing goes on. "It's gonna be a real hard job with a lot of long hours. And I mean *long* hours."

"No," says the recruit, in words more or less like these. "That's what I want to do, get in on the ground floor of a new architecture. I want to do a big machine. I want to be where the action is."

"Well," says Alsing, pulling a long face. "We can only let in the best of this year's graduates. We've already let in some awfully fast people. We'll have to let you know."

("We tell him that we only let in the best. Then we let him in.")

"I don't know," said Alsing, after it was all done. "It was kind of like recruiting for a suicide mission. You're gonna die, but you're gonna die in glory."

4

WALLACH'S
GOLDEN MOMENT

A YOUNG COMPUTER ENGINEER, known to be one of the most skillful in Westborough's basement, said he had a fantasy about a better job than his. In it, he goes to work as a janitor for a computer company whose designs leave much to be desired. There, at night, disguised by mop and broom, he sneaks into the offices of the company's engineers and corrects the designs on their blackboards and desks.

Dreams of pure freedom were not uncommon in the basement. For those who had such fantasies, the best job imaginable would allow them to try to build the unattainable, the perfect computer. What, by contrast, would be one of the worst jobs? One that obliged an engineer to build a kludge. Tom West had to deal with such feelings. This was one of the first and most difficult of his problems.

West needed an architect. In computers, an architecture describes what a machine will look like to the people who are going to write software for it. It tells not how the machine will be built, but what it will do, in detail. Drawing up such a blueprint would be the crucial first technical act in the making of this 32-bit, fully Eclipse-compatible computer that could have no mode bit. West wasn't absolutely certain that such a machine could be made. If it could, what would be the best approach? He had no idea, but he thought he knew who would. Right at the start West decided that a Data General employee named Steve Wallach should be Eagle's architect. "He's the only guy for that job," said West. "The guy's a walking dictionary and encyclopedia of computers. He's the best guy in the world for that job."

Accordingly, West called Wallach to his office in the spring of 1978 and asked him to draw up the architecture for a 32-bit Eclipse.

Steve Wallach glared at West. Wallach got to his feet and, coining a phrase, said: "Fuck that! I'm not puttin' a bag on the side of the Eclipse." Then he stomped out of West's office.

For a time after that, Chuck Holland, an engineer who had been with the team a couple of years, worked on the architecture; he did a great deal of work and an entirely creditable job, as far as he was allowed to go. To West, however, no one else but Wallach would do. He'd get Wallach to sign up somehow. Wallach, he believed, really did want to work on a 32-bit Eclipse, he just didn't know it yet. West knew Wallach. He figured that as much as Wallach wanted to work with a clean sheet of paper and no constraints, he wanted two other things more. These were tangible success and revenge.

Wallach was raised in Brooklyn. His father was a compositor, a practitioner of the craft of hot-metal typesetting, which will soon be all but demolished by the craft that Wallach chose to pursue. There is irony in that, but Wallach didn't think it lamentable. He remembered his father coming home from work with his clothes and hands covered with indelible printer's ink, and his father saying that he did not want his son growing up to come home dirty too. The younger Wallach's talents showed themselves early. He was a frequent and successful entrant in children's science fairs. He went to Stuyvesant High, one of New York City's best public schools; his grade point average — he remembered it exactly — was 93.67 and he graduated forty-eighth in his class. He won scholarships, first to Brooklyn Polytechnic, where he discovered computers and got his bachelor's degree, then to the University of Pennsylvania, where he took a master's in electrical engineering. Then he went to work at Honeywell, in Massachusetts. In one of his first jobs, Wallach worked as junior engineer on a piece of a fancy new computer. But just as that machine was nearly complete, Honeywell merged with General Electric, and when the dust settled the new computer had been scrapped. It never came to light. If Job had been a computer engineer, his travail would have begun in that way.

"Engineers want to produce something," said Wallach. "I didn't go to school for six years just to get a paycheck. I thought that if this is what engineering's all about, the hell with it." He went to night school, to get a master's in business administration. "I was always looking for the buck. I'd get the M.B.A., go back to New York, and make some money," he figured. But he didn't really want to do that. He wanted to build computers.

He took a job at Raytheon, and in time he was assigned to work on another fancy new machine, something called the advanced avionics digital computer; the Navy was paying for its development. According to Wallach, he and another young engineer did most of the actual work on it. One day a team of Navy consultants trooped through their lab. The consultants proclaimed the machine too complex; they said that the engineers would never get it to work. They were wrong. Wallach and his colleague brought the computer to life. "Near the end we probably worked a hundred hours a week," he remembered. "We fought over every little detail, no holds barred, and then we'd go out and play bridge." But then the Navy decided not to build any more of the machines. "To be quite honest, from day one we suspected that the Navy probably wouldn't buy it. It was more that the high-class consultants came in and said we wouldn't get it to work. We said, 'Fuck that, don't tell *us* what we can do.' " All in all, Wallach felt pleased about that piece of work. It was a technical success, he said, and it bolstered his reputation. Once again, however, his machine did not, as West would say, get out the door.

Wallach went to Data General a few years later. The bait that lured him was the promise that he'd get to work right from the start — "on the ground floor" — on the new, supra-state-of-the-art machine that became FHP. Great excitement attended the beginnings of this project. The lucky engineers assigned to it would work on it outside of the plant — in an apartment, for the sake of security, this was such an important project. There was a meeting. An executive tossed a set of keys onto the table. They were the keys to the apartment, and the room number matched the alias of secret agent James Bond: 007. Wallach liked that touch, but he was most impressed by the fact that the company would place no important constraints upon the designers of the new computer. They could pursue pure technical excellence. They were being given a clean sheet of paper. To himself, Wallach said, "Wow!"

He worked on the grand machine for two years. He read and he studied, putting the finishing touches on what was already an encyclopedic knowledge of the best that had been thought and done in computers. Nothing but the best would be good enough for FHP. He lived for that machine. But then the unfinished plans and many of the designers went to North Carolina. Wallach felt he couldn't go. Thus he lost another machine, the best of them all so far.

Wallach was one of several engineers who, working day and night, designed the comely EGO after FHP went south. Then EGO was canceled, for the first time. Wallach got a few things from his office and stomped out of Westborough. He went home for two weeks. Around the time that he returned, DEC announced the VAX. Studying the VAX's architecture, Wallach felt sad. Some of VAX's features were remarkably like EGO's, but EGO was better, Wallach believed. He felt very angry. "We had DEC where we wanted them with EGO."

It was Wallach who suggested to West the idea for Victor. Wallach worked on that machine for a time. Victor died gradually. EGO was revived, and for a little while Wallach really thought that they'd get to build it after all. But de Castro turned it down again.

Wallach had now spent more than a decade working on computing equipment. He'd had a hand in the design of five computers — all good designs, in his opinion. He had worked long hours on all of them. He had put himself into those creatures of metal and silicon. And he had seen only one of them come to functional life, and in that case the customer had decided not to buy the machine.

When EGO-2 got shot down, Wallach went home again, in a rage. Once more, he stayed away for two weeks, but when he came back, he was still angry. He got angrier when West suggested that he create the architecture for a 32-bit Eclipse, because the constraints upon Eagle seemed imprisoning. Eagle would be backward and messy. What a comedown working on it would be!

Clearly, however, Wallach was a man who was ready to get a machine out the door.

The soles of Wallach's cowboy boots faced his office door. He was stretched out in his chair, his feet up on his desk. He was slender, not skinny, with wavy brown hair that was slicked back but still a little unruly. His complexion was pale, for making computers is an indoor occupation and not much vitamin D got into the basement. He was in his mid-thirties.

Wallach had his own, real, if windowless office, down the hall from West's. It was just like West's in its skeleton but differently adorned. Here, enthusiasm lay under less than strict control. Papers lay all over the flat surfaces. Ferns in pots hung from the ceiling. Stuck with pins to the walls were cartoons, T-shirts, posters, postcards and, over by the doorway, a brown paper bag — a joke, the figurative bag on the side of Eclipse given material shape.

One of Wallach's posters was a blown-up reproduction of Saul Steinberg's droll cartoon depicting the United States: New York

City occupies the entire foreground of the picture; California is awarded a much smaller, but significant, space in the background; and the states in between take up about as much room as the lettuce in a club sandwich. Wallach said he liked this poster for its accuracy. It also made him think of North Carolina, one of those states that got short shrift in the cartoon. Wallach had been offered the chance to go south with FHP. Indeed, he had been encouraged to go. Why had he refused?

"I thought North Carolina sucked," he explained. "I told them, 'That's not a place where I want to bring up my wife and family.' I was given a tour of Chapel Hill and somebody told me that more people per capita read the *New York Times* in Chapel Hill than in all the rest of the state. I said, 'It sounds like the Warsaw Ghetto you just described.' Hey, when I wake up in the morning I want to hear the stock market report on the radio, not the price of tobacco and hogs. And what they considered good food — I wouldn't wish it on my worst enemy." Wallach sniffs, wrinkles his nose, and then delivers such lines with his chin forward, as if daring someone to hit it. "I said, 'That's not a place where I want to bring up my wife and children.' " He made me laugh. He watched, smiling.

Carl Alsing gleefully related that once, when an engineer in the basement was describing how rough it had been to grow up in India, Wallach burst in declaiming that India was easy — that man should have tried growing up in Brooklyn, where you had to fight your way from one ethnic neighborhood to another just to get to school.

Wallach remembered that when he was very young, another boy on his street punched him and he didn't hit back. His father punished him for this failure to retaliate. Asked to describe his old neighborhood, Wallach said: "Let's put it this way. You learn very quickly that if someone hits you, you hit them back twice as hard, so they won't hit you again."

When I met Wallach, he was living in a residential neighborhood in Framingham, Massachusetts, in a new and prosperous-looking housing development. His house was gray and clap-boarded in the Colonial style. It was pretty. Inside, it was immacu-late. He sat in his living room on a white easy chair telling what amounted to old war stories. One concerned an executive whom Wallach had believed to be an enemy of the Eclipse Group. Wal-lach had decided to "play with the turkey's mind." For a week, everytime he ran into this man, Wallach would smile at him. The following week, however, whenever Wallach encountered the man, he would make himself look sad. Finally, the executive could stand it no longer and asked for an explanation — which was what Wallach had been waiting for.

"Steve, how come you're so happy to see me sometimes, and then sometimes you look like you wish I wasn't there?"

"I don't know," said Wallach. "I'm very moody."

"Then," recalled Wallach, "I went into my office, closed the door, and got hysterical."

Wallach's wife overheard this tale. "Steve, that's terrible," she said. "I live with this," she added, turning to me. But I noticed that she was smiling.

While Wallach was talking in this vein, his four-year-old daughter was circling the room on roller skates, across the wall-to-wall carpeting. Considering the whiteness of his chair and his proclivity to battle, I expected Wallach to yell at her at any mo-ment, but I soon got the feeling that he did not often raise his voice in anger while at home. When the little girl skated up be-hind him and planted a wet, loud kiss on his cheek, Wallach beamed. "That's why you have daughters."

You had to understand the context in which Wallach practiced his martial arts. His swords were usually technical issues, and he left them at work. Moreover, in the last few years, he had been sorely provoked. He felt, he said, "like an animal thrown in a cage."

In this situation, Wallach allowed, he was a rough customer. On one of his many visits to North Carolina during the so-called

EGO wars, he was presented with a poster depicting the hideous face of the arch villain of *Star Wars,* Darth Vader himself. "We got this just for you," Wallach was told. He liked to tell that story.

When FHP's removal south had been announced, Wallach had felt robbed and insulted. Then he had felt astonished and dismayed when he had read the article in the *Globe* that seemed to have de Castro denigrating the engineers at Westborough. Wallach called up the reporter. Pretending to be an important Data General executive, he badgered the fellow into reading his notes over the phone. On the basis of this conversation, Wallach decided that de Castro had been quoted out of context. But the story left a bad taste. Wallach heard the people who were going south with FHP say that North Carolina was where the action was and would be from now on. He became convinced that neither the defectors to North Carolina nor de Castro himself believed that the stay-behinds at Westborough would produce significant machines in the future. He took the whole affair as a personal insult. "A slap in the face." He had to retaliate. Working on EGO was his first attempt to do so. "Like I said, I'm a fighter, and the challenge was to prove the man in the corner office and North Carolina wrong."

In the EGO wars, Wallach served as West's Hessian. West didn't feel angry toward North Carolina, nor could he afford to appear hostile. But arguments are arguments; certain nasty things had to be said. Wallach was good at saying them and happy to do it. He served, he said afterward, as "West's gun" in the "shootouts."

But in the end Wallach got beaten. EGO was canceled — not because it wasn't a good design, but because it was too good, too much of a threat to steal the show from North Carolina. West never fully believed this, but Wallach always did.

When West started talking, after all of that, about building a 32-bit Eclipse, Wallach sneered — partly because it seemed to him that this was just the sort of project that North Carolina would have wished on him: an insignificant, kludgey machine that probably wouldn't get out the door. He'd rather stand in honorable defiance than work on such a thing. He was quitting the fight, he said, and soon he would leave Data General.

Meanwhile, thinking in the fastness of his office, West decided that he had made a mistake. He had not presented Eagle in the proper light to Wallach. Over the next few weeks, he had many long conversations with the reluctant architect. He listened to Wallach's complaints. He agreed that Wallach had been treated badly.

"But haven't you realized yet that the way to prove someone wrong is to build the right thing?" West asked, time and time again. Finally, somewhat exasperated, but also suspecting that Wallach was ready to yield, West forced the issue. "Either you do this or your job description is inoperative," he said to Wallach one day.

The phrasing appealed greatly to Wallach, and he could see that West was probably right. Conceiving architectures was his job and Eagle was the only project around that needed an architect. But Wallach had seen too many projects canceled in spite of their merits to believe that Eagle would go out the door just because it promised to be a good commercial machine. He wasn't going that route again. Wallach wanted to talk to de Castro. This was not a completely extraordinary request; de Castro's office, Wallach knew, would usually open for an engineer who had something to say.

As Wallach remembered it later, for those who questioned him about it, this was the gist of his conversation with the president:

"Can I be straight with you?" asks Wallach.

De Castro nods.

"Okay," says Wallach. "What the fuck do you want?"

"I want a thirty-two-bit Eclipse," says de Castro.

"Are you sure? If we do this, you won't cancel it on us? You'll leave us alone?"

"That's what I want, a thirty-two-bit Eclipse and no mode bit."

Wallach returned to West's office, and now, at long last and sniffing, he said, "Okay, Tom, one more time."

"The document's yours," West replied. "You gotta do it fast."

Wallach went to his office and closed the door. Some months later, a careful examination of Wallach's quarters revealed many scuff marks etched in shoe polish low down along his walls, and there was a dent higher up on one wall. These were bruises, left over from Wallach's labor on the machine.

When he sat down, alone, in his office, Wallach reasoned that since the whole purpose of this ridiculous undertaking was 32-bit-hood — the enlargement of the Eclipse's logical-address space from 65,000 to 4.3 billion storage compartments — he might as well begin by figuring out how the compartments would be organized ("managed") and the information in them protected. He further decided — he called this "the methodical engineering approach" — to worry about memory management first. Clearing a space on his desk, he placed a yellow legal pad in front of him and drew a picture of the standard 32-bit address — a box containing 32-bits, which looked like this:

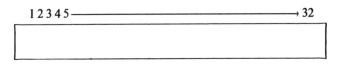

He began to divide up the space inside this box.

If you imagine the computer's storage — its memory — to be a large collection of telephones, then what Wallach was doing might be described as designing a logical system by which phones and groups of phones could be easily identified — a system of area codes, for instance.

Next to Wallach's desk stood several tall metal bookcases. Their shelves from floor to ceiling were stuffed, with loose-leaf binders and fat volumes in plain covers that bore titles such as

Parallelism in Hardware and Software: Real and Apparent Concurrency. The binders contained specifications for nearly every sort of computer ever made and for some that had never been constructed. Wallach called his bookcase "Data General's de facto library." He claimed that he had most of this library in his head. From time to time, he wheeled around in his chair and took a binder down from the bookcase. By the end of the day, Wallach had roughed out the divisions of the standard 32-bit address, and he said to himself: "Okay, great. So far I've done nothing." He gave the wall near his door a kick as he left.

Wallach wasn't ready to admit that he was having fun. But he was back in his office early the next morning. He had a workable general plan for managing memory. Now for the question of how to protect the stored information.

This had become one of the important questions in the computer industry. It had arisen largely because of a practice called time-sharing. In the basement of Westborough, for example, every engineer had a computer terminal — typewriterlike keyboard plus video screen — and most of these terminals were hooked up to a single large Eclipse, located some distance away, behind one of the locked doors that warned against unauthorized entry. So long as the central computer didn't become overloaded, each of the engineers, at his separate desk, could enjoy the illusion that he alone had access to the computer, though in fact they were all sharing it. It was a common sort of system. Much more exotic ones existed. Time-sharing already took place not only within buildings, but between them and across continents and oceans. Increasingly, computers communicated with other computers across vast distances.

Many organizations stored very valuable information in their computers. In a sense, banks store their money in computer systems, and oil companies store their crown jewels, their seismic information, in computerized data banks. Publicized cases of thievery and espionage by computer were legion at this time; experts generally agreed that most electronic criminals were never caught and that many who were never got prosecuted, because an organization that had been successfully attacked either didn't want the embarrassing publicity or feared that the news would encourage other raids.

Many people had taken a crack at solving such problems, notably a group of engineers and computer scientists at MIT who worked with money from the Department of Defense on a project called MULTICS. In the late sixties, they produced a complex plan for making time-sharing systems secure. It was a clever plan. But many experts believed that no system of protection yet devised could withstand the efforts of smart pranksters or thieves bent on foiling it. One organization had purchased a very fine system for protecting its computerized data banks, but a determined group had cracked it. Pretending to be the manufacturer of the security system, they had sent to the organization a bogus set of revisions to the system's software. Without affecting the system's performance, these revisions left open a "trapdoor," through which the thieves were able to withdraw important information from the organization's data bank. The anecdote was one of Wallach's favorites. "I'm not gonna try and solve the world's problems here," he conceded. He'd forget about trying to stymie the malicious user and concentrate instead on preventing accidental damage.

Inadvertently, users of a time-sharing system could alter the contents of the host computer's memory and in this way destroy valuable data and foul up system software. Plans such as MULTICS were workable solutions to this problem.

Wallach was proceeding almost on instinct now. In his two years of work on FHP, he believed, he had read every published description of every system for protection that had ever been devised. In the ordinary case, an architect might linger over the various possibilities for months. Wallach didn't have time for that, and he really didn't need it. Very quickly, he chose what he believed to be the simplest and best general solution, the general idea that came out of MULTICS — which DEC, as it happened, had used in the VAX. This was a system of "rings."

Picture an Army encampment in which all the tents are arranged in several concentric rings. The general's tent lies at the center, and he can move freely from one ring of tents to another. In the next ring out from the center live the colonels, say, and they can move from their ring into any outer ring as they please, but they can't intrude on the general's ring without his dispensation. The same rules apply all the way to the outermost ring, where the privates reside. They have no special privileges; they can't move into any ring inside their own without permission.

Analogously, each ring defines some part of the computer's memory. How does the computer mediate a user's access to these various areas? In the general case, it compares two numbers — the number of the ring to which the user has free access and the number of the ring that the user wants to get into. If the user's ring number is smaller than or equal to the number of the ring that the user wants to enter, the user is allowed to go in. But how should ring numbers be assigned to the compartments of memory? That was the question before Wallach.

With the VAX, DEC's engineers had solved the problems of memory management and the protection of memory separately. Each compartment in memory had an address, and each compartment also had a separate ring number, which was produced and checked by a special set of hardware. Wallach had studied the specs on VAX. He didn't like that approach. From the back of his mind came the recollection of a conversation he'd had at a conference of engineers from different companies some years before. At such convocations the usual way of getting acquainted is to ask other engineers what projects they've worked on. Wallach remembered an engineer telling him about a ring system he'd thought up and never built, in which ring numbers and addresses were mingled. Wallach had gotten the fellow to send him a spec; it had looked a little clumsy to him. But there was the germ of something there. Wallach drew another diagram of the standard 32-bit address:

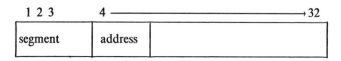

The first three bits of the address would contain the segment number of a memory compartment — in the telephone analogy, a given compartment's area code. The other bits would define the rest of the address; but Wallach wasn't interested in them just now. He was pondering the first three bits. Suddenly, without thinking about it, he was drawing another box below the first box. The diagram now looked like this:

1 2 3	4 ⟶ 32	
segment	address	
ring number	address	

The segment number — the area code — would be the same as the ring number, which defined the level of security to which the compartment would be assigned. Three bits can be combined in eight different ways. So there would be eight rings (eight levels of security) and eight segments (eight area codes) in the memory system. The area codes would themselves indicate which ring was forbidden to whom.

Although they are generally shy about claiming to have had one, engineers often speak of "the golden moment" in order to describe the feeling — it comes rarely enough — when the scales fall from a designer's eyes and a problem's right solution is suddenly there. The chief virtue of Wallach's scheme was its simplicity. It would be relatively cheap and easy to implement in hardware and software, and it should work efficiently and reliably. When Alsing saw Wallach's brief description of the plan, he said to Wallach, "That's nice." Later, out of Wallach's earshot, he said more. "Rings have been around. They're old hat. What makes Wallach a good Data General engineer is that he came up with a really elegant subset of those ideas — simple, sweet, cheap, efficient, clean. And I can't believe I just said that about Wallach."

As for Wallach, after he had drawn the diagram, he stared at it, wondering for a moment, "Where did that come from?"

He kept eyeing it. "That looks pretty cool."

For the next couple of days Wallach played with the idea, until he was sure it would work. Then he pulled his computer terminal in front of him and wrote up a memo, which told how others had handled memory management and protection in the past and which described his own general scheme. Near the end of the memo, his sarcasm rose. He described the plan as a kludge, but a better one than DEC had put into VAX. He kicked his walls a few times. The idea of placing that neat, clean structure on top of the outdated structure of the Eclipse repelled him. It was as if he had invented a particularly nice kind of arch for the doorway of a supermarket.

Wallach was a scholar of computer architectures. He knew by heart the works of the Michelangelos, the Frank Lloyd Wrights and the Gaudis of the computer profession. He imagined himself standing in front of a roomful of experts when Eagle was all done. They would question him about the architecture, inserting sharp little knives in his flank and then twisting them. "Why didn't you do it this way, Steve, when it's obviously better?" And, as he imagined it, his only possible defense would be to plead weakly and traitorously that the company had prevented him from really showing his stuff.

The cause of this fretting lay mainly with the "instruction set." These are the basic operations that the computer's builders equip it to perform. Typically, instructions bear such names as ADD, which means the computer should perform addition, and Skip On

Equal, which tells the machine to compare two values and if they are equal to skip the next step in a program. Today, there are a couple hundred instructions or so in most minicomputers' sets. A large part of the art of designing a computer's architecture lies in selecting just the right set of instructions and in making each instruction as versatile as possible. This art had advanced a great deal since the invention of the Eclipse. But in order to be fully compatible with Eclipses, Eagle would have to contain the same old instruction set in its entirety. DEC had not tried to do that. VAX was not fully compatible with DEC's old 16-bit machines. DEC had given up full compatibility for the sake of giving VAX what Wallach called a wonderful — a "super" — instruction set. He was somewhat biased. EGO's instruction set had resembled VAX's. Both were examples of the state of the art. The Eclipse's instruction set was an example of what that state used to be.

For many months, Wallach would continue to mourn EGO outwardly and to tell anyone who was interested that if some expert asked him someday why he hadn't invented a better instruction set, he would spill the beans. He'd tell them he had been forbidden to use a mode bit. "I'll say it was because I was told that these were management's objectives and I was told that I couldn't work on the machine if I wouldn't fit in with management's objectives." But in his own mind, he was changing his tune. He was getting to like the looks of this architecture. He was starting to think of it not as a wart on a wart, but as a clean design with a wart on it. The wart was the Eclipse instruction set, virtually every part of which Eagle would have to contain, for the sake of compatibility. But there were some other empty corners of this canvas, aside from memory management and protection — chiefly, the new 32-bit Eagle instructions. Wallach came up with some that he liked a great deal. In fact, he even found a way to slip in a well-disguised equivalent of a mode bit — which would have allowed him to define a wholly new set of Eagle instructions, ones not at all derivative of Eclipse. But he didn't disguise the mode bit well enough. West found him out. "We're not gonna do that," he said to Wallach. Wallach went back to his office and punched the wall near his door.

But West did let his adjutant put into Eagle some new instructions that weren't Eclipse-like. Not all the ideas for new instructions came from Wallach, but sometimes it worked this way. They developed a routine. Wallach would bring West an idea for a new instruction. West would say that it looked like a win, but that it wasn't Eclipse-like. Wallach knew what that meant. If the wrong people saw this new instruction in the spec, it might cause a stir. People upstairs and in North Carolina might get the idea that Eagle was, after all, intended as a challenger to FHP — which it certainly was. So Wallach would take the idea for the new instruction to friends working in System Software, and the system programmers would approve the idea. They and Wallach would sit down and fully define the new, non-Eclipse-like instruction, and then Wallach would ask them to write a memo to the Eclipse Group, requesting that this instruction be put into Eagle. *"They* wrote the memo," said Wallach, "so that the idea would be perceived as coming from them, just in case we ever got called on it." Wallach went on: "A lot of things we did were unique to that environment. It's clear they weren't always the way things should be done."

"But you enjoyed doing things that way?" I suggested.

"We all enjoyed it," he said. "Anytime you do anything on the sly, it's always more interesting than if you do it up front."

After his discovery of the basic scheme for memory management and protection, Wallach had to work out the details, in many long and sometimes loud arguments with Ken Holberger, sublieutenant of the Hardy Boys, who were going to have to implement the architecture. For technical argument, Wallach found Holberger to be a worthy adversary, and on the whole he enjoyed those debates.

Next, all of the instructions had to be defined, as did the precise mechanism by which Eagle would move smoothly, and without the intervention of the user, from programs written for 16-bit

Eclipses to ones made for a 32-bit machine. This was a tricky, time-consuming piece of work. Wallach also had to collect all these details and schemes in a document. He took great pains with this volume. He called it "my book," and refining it, rewrote it seven times over the following months. This book was some two hundred pages long, and at the beginnings of each chapter he placed a famous or semifamous quotation.

Wallach said he never read technical tracts outside of work and that he shared West's suspicions about people who did. At home, he said, he mainly read *Playboy.* "I read the short stories. I really do! Yeah, I look at the pictures, but I like the stories." He laughed. He had not read widely in the classics. So the epigraphs for his book didn't come to him without effort. He took the quotes from Victor Hugo, Nietzsche, Shakespeare, T. S. Eliot, Santayana and FDR. Some were playful and some downright witty, if you understood the context. At the top of the chapter about the instruction set, for instance, he placed this quote from *Macbeth:*

> *We still have judgement here; that we but teach*
> *Bloody Instructions* [Wallach's cap.], *which, being taught, return*
> *To plague the inventor.*

For the chapter about his elegant scheme of memory management, he chose these verses from Tolkien's *Lord of the Rings:*

> *It cannot be seen, cannot be felt*
> *Cannot be heard, cannot be smelt,*
> *It lies behind stars and under hills*
> *And empty holes it fills.*

Wallach spent about twenty hours in the Framingham Public Library, with his nose in *Bartlett's Familiar Quotations* and dipping into some of the actual works, just in order to add these flourishes to his spec. They added something. They revealed the class of feelings that Wallach brought to his job. If he was a Hessian, he was a passionate one, and with the quotations he signed his name to his piece of the new computer. Wallach actually spent far more time looking up epigraphs than it took him to discover the right way to manage and protect the computer's memory. But that small golden moment colored everything else for him. As he saw it, the rest of the plan simply unfolded from that one idea. It was a good omen.

How do such moments occur? "Hey," Wallach said, "no one knows how that works." He remembered that during the time when he was working on the Navy computer for Raytheon — the one that got built and then scrapped — he was at a wedding and the solution to a different sort of problem popped into his mind. He wrote it down quickly on the cover of a matchbook. "I will be constantly chugging away in my mind," he explained, "making an exhaustive search of my data bank."

5

MIDNIGHT PROGRAMMER

IT WAS THE HOUR of insomniacs. In the basement of Westborough, the corridors and cubicles stood empty and in shadows. Carl Alsing's cluttered little area made a small rectangle of light. Strewn before me across the surface of his desk, like the relics of a party, lay dozens of roughly drawn maps. They consisted of circles, inside of which were scrawled names such as Dirty Passage, Hall of Mists, Hall of the Mountain King, Complex Junction, Splendid Chamber, Bedquilt, and Witts End. Webs of lines connected the circles, and each line was labeled, some with points of the compass, some with the words *up* and *down*. Here and there on the maps were notations — "water here," "oil here," and "damn that pirate!" In the midst of all this paper sat Alsing's computer terminal. On the screen of the tube in white letters, like the little voice that whispers in a wild gambler's ear, this message stood:

ARE YOU SURE YOU WANT TO QUIT NOW?

Alsing had offered to demonstrate "midnight programming," and now, early in the winter of 1979, he had made good on the promise. He was sitting beside his desk with his hands folded, wearing his soft, sneaky smile. For hours, he had been watching me play the game Adventure. "You really got into it," he said. "That's good."

When first invented, the program for Adventure had traveled widely, like a chain letter, from coast to coast among computer engineers and buffs. It had arrived in Westborough just in time for the aftermath of the EGO wars. It was everywhere by now; grade-schoolers were playing it.

In the game, the computer appears to create for you an underground world, called Colossal Cave. It moves you through it in response to your commands. The computer seems to act as game board, rule keeper and, when you foul up, as both assistant and adversary. You move by typing out directions on the keyboard at your terminal. If you spell out these directions in full a few times, a message will appear on your screen, saying:

IF YOU PREFER SIMPLY TYPE N RATHER
THAN NORTH

"How did the computer know to do that?"

"I don't know," said Alsing, coyly. "Sometimes it's perceptive, other times just dumb."

After you have moved, a message appears on the screen telling you where you are and what you are confronting. You must respond, in two words or less, both to opportunities — treasure or tools lying on the floor of some chamber — and to threats and challenges — the hatchet-hurling dwarf, the snake, the troll who guards the bridge, the dragon. If, for instance, you want to get past the rusty door in one of the chambers, you have to think of what will conquer rust, then you have to remember where it was you saw that pool of oil, then you have to type in step-by-step instructions to get back to that oil, and then, because the computer will let you carry only so many things, you may have to drop one of your tools or treasures — DROP GOLD COINS, you might write — and then type in, TAKE OIL. Of course, you must already be holding a container for the oil. Then you have to retrace your steps back to the rusty door and type, OIL DOOR. This method of travel and maneuver takes some getting used to, but after a while it's as easy as driving a car. The game is a harrowing of Hell.

Earlier that night, traveling more or less at random and without maps, I had stumbled into a place that the message on the screen described this way:

YOU ARE IN A MAZE OF TWISTY LITTLE
PASSAGES, ALL DIFFERENT

So I typed, NW, in a hurry, hoping to get out of there by going back the way I'd come. But the screen responded:

YOU ARE IN A TWISTY MAZE OF LITTLE
PASSAGES, ALL DIFFERENT

If you are susceptible to Adventure, you get worried at this point. I had the feeling I was lost in a forest, and I acted as no smart woodsman would, heading off in this direction, then heading off in that, and getting nowhere.

Then I heard Alsing chuckle. "Ahhh, I love it."

I thrashed around in that maze awhile longer. Finally, Alsing said, "Look carefully at the messages on the screen."

"They're all the same."

"No, they're not."

Each chamber of this maze within the big labyrinth had a slightly different and unique address, formed by a particular arrangement of the words *twisty, little, passages,* and *maze.*

"And what do you do when you get lost?" asked Alsing.

"You make maps, of course."

Alsing sat back and nodded, smiling — the complacent schoolmaster, one of the roles he played at the beginning of Eagle.

Later on, though, I wandered into a maze that really scared me.

YOU ARE IN A MAZE OF TWISTY LITTLE PASSAGES, ALL ALIKE

You have to find your way around this maze if you hope to begin to master Adventure, because this one contains the vending machine with the batteries for your indispensable flashlight and, moreover, harbors the lair of the kleptomaniacal pirate who is forever sneaking up behind you and snatching away your treasures. But how do you find your way around a maze in which all the chambers are the same? You must do what Hansel and Gretel tried to do, more or less — drop something on the floor of every chamber to leave a trail marking where you've been.

That isn't the worst maze, however. You can get caught in Witts End and think that you'll never get out. Some of the engineers at Westborough who had come close to mastering the entire game believed that the only way out of Witts End was to tell the computer you want to commit suicide — AXE ME. That works; you get reincarnated shortly afterward. But you lose points; suicide isn't the best solution.

I myself did not get as far as Witts End, however, but quit while in the maze that was all alike. And though the computer seemed reluctant to let me go — DO YOU REALLY WANT TO QUIT NOW? — I stood firm. Alsing got up and led the way toward the cafeteria. He got us lost in the corridors of the basement. No doubt that was his plan, though he denied it. Intentional or not, our getting lost allowed him to cry out, "The twisty little passages of Data General!"

It was the time of night when the odd feeling of not being quite in focus comes and goes, and all things are mysterious. I resisted this feeling. It seemed worth remembering that Adventure is just a program, a series of step-by-step commands stored in electrical code inside the computer.

How can the machine perform its tricks? The general answer lies in the fact that computers can follow conditional instructions. They can take two values and compare them — that comes down to simple arithmetic — and, if so commanded, can perform one action if the values are equal and another if they are not. In this ability to follow conditional instructions — an ability built into the machine — lies much of the computer's power. You can set before it, in sequence, bifurcating webs of conditional instructions, until the machine appears to make sophisticated decisions on its own.

When we had returned from our adventure finding coffee, I asked Alsing how he felt about the question — twenty years old now and really unresolved — of whether or not it's theoretically possible to imbue a computer with intelligence — to create in a machine, as they say, artificial intelligence. Alsing stepped around the question. "Artificial intelligence takes you away from your own trip. What you want to do is look at the wheels of the machine and if you like them, have fun."

To the two computers that the Eclipse Group used, the engineers had given the names Woodstock and Trixie, after characters in comic strips. They often spoke about these computers as if they had personalities. When especially frustrated, one Microkid would walk into the lab where Trixie resided and yell at the machine. Alsing said: "A lot of people are really tired of anthropomorphizing computers, but it sure is an easy way to talk about them. You can anthropomorphize your car and the analogy works, and then at some level it doesn't. We anthropomorphize big business, the military and so on, as some strange creatures with alien personalities. I think that's sane, I think that's normal. You tend to have to anthropomorphize the computer. It presents a face, a person to me — a person in a thousand different ways."

He drew his chair up to his terminal and typed a few letters — a short code that put him in touch with Trixie, which was the machine reserved for the use of his microcoding team. "We've anthropomorphized Trixie to a ridiculous extent," he said.

He typed, WHO.

On the dark-blue screen of the cathode-ray tube, with alacrity, an answer appeared: CARL.

WHERE, typed Alsing.
IN THE ROAD, WHERE ELSE! Trixie replied.
HOW.
ERROR, read the message on the screen.
"Oh, yeah, I forgot," said Alsing, and he typed, PLEASE
HOW.
THAT'S FOR US TO KNOW AND YOU TO FIND OUT.
Alsing seemed satisfied with that, and he typed, WHEN.
RIGHT FUCKING NOW, wrote the machine.
WHY, wrote Alsing.
BECAUSE WE LIKE TO CARL.
One of Alsing's Microkids had programmed Trixie to deliver
these impertinent responses. In a real sense, Alsing was conversing
with a member of his team. I think — one of them said as
much — that if the Microkids' computer had ever started talking
back to them all on its own, they would have ripped its wires out.

Anthropomorphizing Trixie was just a game, and by no means
the most sophisticated that Alsing played with his team. They
were the imps of the Eclipse Group and their computer showed it.

Assembled from the comments of a few of the young men on
the team, the typical engineer, an imaginary creature perhaps,
wears a white undershirt and a plastic pouch (known to some as a
"nerd pack") in his breast pocket, in order to keep his pens from
soiling his clothes. An electronic calculator — used to be a slide
rule — hangs like a ring of janitor's keys from his belt. Jim Guyer
of the Hardy Boys, who wore a beard and drove a large motorcycle, added, "A lot of people think that an engineer is shut off in
a little corner and doesn't give a damn about anything else except
his own little thing," Guyer said. "They exist. And they're
the most obvious engineers, because in their isolation they're
obvious."

When first encountered, Alsing might seem to fit this description. He could be taken for one accustomed to dark corners. In
this respect, as in others, he is deceiving.

Alsing is tall, over six feet, but he doesn't seem to realize it and he doesn't look that big. He is neither fat nor thin. He keeps his hair cut fairly short. Often his dress looks sloppy — not deliberately or extremely so, but slightly careless. He speaks softly, as a rule, and the pitch of his voice, though not at all squeaky, is high. About his hands, the way he folds them in his lap or puts them together under his chin, there is something delicate. One acquaintance of his said, "He looks uncoordinated." In fact, Alsing does not play many physical games, though on a vacation in the Caribbean he took up scuba diving. Ham radio is an old hobby of his. I thought that I could see in him the lonely childhood behind him — he would have been the last boy picked in every schoolyard game, the one who threw a ball like a girl. But Alsing is gregarious. He set out some years back to master social gracefulness, and it shows. He will sit down in a strange living room, fold his hands neatly in his lap, and listen. You can forget that he is there, until gradually, so delicately you hardly notice, he enters the conversation. After several such occasions, people who had just met Alsing said to me, "He's really smart and interesting, isn't he?"

By the time the Eagle project began, Alsing had attained the age of thirty-five, which made him very elderly within the group — but he was a clever and playful old codger. His eyes would dart around. His eyebrows would dance; they'd do his winking for him. When he'd close his mouth and let a little smile creep across his cheeks — in each one was carved a line in the shape of a half-moon — then it was time to be on guard, for a game, a trick or a joke.

Alsing's childhood did not leave him with an abundance of sweet memories. Discovering the telephone was one of his fondest. When he was eight and in third grade, his family moved from central Massachusetts to Evansville, Indiana. He loathed that place. "I was smaller, paler, weaker, less rugged and I had a funny accent. I remember discovering in third grade that there was a pecking order — 'Hey, there's a pile and I'm on the bottom of it.' I was a very depressed kid in third grade." He vividly recalled the day when he skipped recess, usually a painful event for him anyway, and instead worked at his desk on the design of a telephone. He wanted to find out how the thing could possibly capture a voice. It seemed to him an improbable instrument, one that shouldn't work. He read about telephones in several encyclopedias. Then he took the family phone apart. Finally, he figured it out to his satisfaction. "This was a fantastic high, something I could get absorbed in and forget that I had these other social problems."

One day, while at home in Evansville, he was prowling around the basement and noticed wires running across the ceiling of the coal bin. He traced them back to the phone upstairs. He got ahold of some batteries, an old microphone and an ancient set of earphones, and from the coal bin — sitting there all alone in the grime, with the earphones on — he tapped the family's telephone. Once in a while, he accidentally short-circuited it, but his father, an engineer for Westinghouse who designed refrigerators, was indulgent on that score.

Some years later, Alsing's family returned to New England, and he enrolled at the University of Massachusetts. He made mostly poor marks. Then, in his junior year, he took a course in the theory of digital circuits. It entailed, as such courses always do, the study of Boolean algebra. Alsing's world was never the same after that. Fathoming this algebra, Alsing felt as some children do when all at once they know how to read. Boolean algebra was something that made perfect sense, and thus it was a rare commodity for him. He called it beautiful.

At the heart of the computer lies a device made up largely of transistors. Engineers call this device a gate, and the analogy is apt. You might think of it as a newfangled, automatic barnyard gate in an electrical fence. If the gate is closed, current flows through it along the length of the fence; but the current stops at the gate when it stands open. You open and close this gate by sending signals to it down two or more (let's say it's two) control wires.

How can you add two numbers together with electricity? For Alsing, the question had the force of those that the telephone had long ago aroused in him. You assumed, first of all, that you were going to do your adding in binary arithmetic. It was simple, once you got the hang of it. You can count as high as you like in binary, but you use only the integers 0 and 1. The zero of the familiar decimal arithmetic is 0 in binary, and the one in decimal is also 1 in binary; but two in decimal is 10 in binary, three is 11, and so on.

Next, you let a high voltage represent the binary integer 1 and a low voltage represent the binary integer 0. Then you build a gate, which for all practical purposes is a binary device: it's either opened or closed. When opened it passes on a low voltage, which stands for 0; closed, it passes on the symbol for 1. What's crucial is how this gate responds to signals from the two wires that control it. You can build your gate in such a way that it will close and pass on the electrical symbol for a 1 if, and only if, one of its control wires sends it a 1 and the other sends it a 0. A gate built this way thus will add 0 and 1 and yield the right answer every time.

But if you want to add 1 and 1 and get the right answer, which is binary 10, then you have to modify your circuit, and if you want to add large numbers together, you have to build a large array of gates and control wires. Suppose you want to build an adder that can combine two 32-bit packets. The diagram for such a circuit looks forbiddingly complex. How can you be sure that for every possible set of signals you put into this circuit, it will produce just the right pattern of opened and closed gates to yield up the right answer? You need a set of rules. Boolean algebra, for instance.

Conventional algebra sets rules about the relationships between numbers. Boolean algebra expresses relationships between statements. It is a system of logic; it sets general conditions under which combinations of statements are either true or false. In this sense, it is a binary system. What Alsing studied was in fact a simplified form of Boolean algebra, one that had been tailored to digital circuits. "If, and only if, A is true AND B is true, then their combination is also true" — that is one of the statements of the algebra. There are others, and it turns out that gates can be built so that they behave precisely according to these statements. Indeed, the kinds of gates are named for the Boolean statements that they mimic; there are, among others, AND gates, NAND gates, OR gates, NOR gates. Boolean algebra provides one systematic way for engineers to design their circuits. It is less important for that purpose than formerly, but it was a crucial tool in Alsing's college days.

The time was the middle sixties. The computer was a famous and closely held marvel; the day had not arrived when curious undergraduates could buy the parts and figure out the thing by building one themselves. But here, in the course that Alsing took, lay the secret of the machine. "It was so simple, so elegant," he said. He got an A in that course. So he took a course in how to program a computer by using a high-level computer language called FORTRAN, which was developed for scientists mainly. Alsing got an A in that. He flunked everything else — because of an actual computer.

It was an IBM machine, archaic now but gaudy then. The university owned it, in effect, and it lay inside a room that none but the machine's professional caretakers could enter during the day. But Alsing found out that a student could just walk into that room at night and play with the computer. Alsing didn't drink much and he never took any other drugs. "I was a midnight programmer," he confessed.

During the first nights after he learned to write a computer program, Alsing would go off from the computer room and search the empty building, looking for a classroom with a blackboard and some chalk. He posed problems for himself and on the blackboard wrote up little programs for their automatic solution. Then he hurried back to the computer to try out his programs on the machine. That was what made it fun; he could actually touch the machine and make it obey him. "I'd run a little program and when it worked, I'd get a little high, and then I'd do another. It was neat. I loved writing programs. I could control the machine. I could make it express my own thoughts. It was an expansion of the mind to have a computer."

About ten other young, male undergraduates regularly attended these sessions of midnight programming. "It was a whole subculture. It's been popularized now, but it was a secret cult in my days," said Alsing. "The game of programming — and it is a game — was so fascinating. We'd stay up all night and experience it. It really is like a drug, I think." A few of his fellow midnight programmers began to ignore their girlfriends and eventually lost them for the sake of playing with the machine all night. Some started sleeping days and missed all their classes, thereby ruining their grades. Alsing and a few others flunked out of school.

In a book called *Computer Power and Human Reason,* a professor of computer science at MIT named Joseph Weizenbaum writes of a malady he calls "the compulsion to program." He describes the afflicted as "bright young men of disheveled appearance, often with sunken, glowing eyes," who play out "megalomaniacal fantasies of omnipotence" at computer consoles; they sit at their machines, he writes, "their arms tensed and waiting to fire their fingers, already poised to strike, at the buttons and keys on which their attention seems to be as riveted as a gambler's on the rolling dice." Was this a portrait of Alsing playing with that old IBM?

In order to explain the hold that machine once had on him, Alsing had insisted that we confront Trixie late at night. As if preparing for a journey or an execution, we had dined well that evening. And when we'd gotten to the basement, he'd decided to start me on the way to addiction with small stuff — that is, with Adventure. I'd played and played. One by one the night owls of the basement — and there were many in the Eclipse Group then — had departed. I'd kept on playing.

When I finally quit, I felt weary in my bones. I was actually sweating; my shirt stuck to my back. Things around me kept going in and out of focus. I looked at Alsing, and the rims of his eyes were red. He said he could remember experiencing weariness like this during his midnight-programming days, but he had been younger then. Weariness had been a badge and part of the fun. Some of his cohorts had suffered, it was true. "But college kids are vulnerable. They can get taken down by girls, or drink, or by programming." As for him, he felt that he had gained far more than he had lost.

Up until that year when he discovered the machine, his life, every way he had looked at it, had seemed chaotic. He had done fairly well in a course in psychology, but in nothing else. He had proven to himself that he was an inveterate failure. Now, finally, he felt truly interested in something. He left college for a year. When he came back, he took a number of courses in electrical engineering and became a straight-A student. He took a job with DEC when he graduated, then went to Data General, partly because he thought it would be a lively place to work; he got that idea from the famous first ad and from the angry talk about Data General he heard in the hallways at DEC.

At Data General, the number of your security badge reflects your longevity: the lower the number, the longer your service. By 1979 new employees were receiving badges numbered in five figures. Alsing wore number 150, which was low enough to confer upon him some status and, he thought, perhaps a measure of security. According to several stories, de Castro had shown a special loyalty toward those who had joined his company early on and stayed. At Data General, Alsing became a microcoder. In fact, he was Data General's first and most prolific one.

Most new microcoders, on their first job, have the odd feeling that what they're doing can't possibly be real. "I didn't fully believe, until I saw it work, that microcode wasn't just a lie," said Alsing's main submanager, Chuck Holland, remembering the first code that he wrote. At the level of the microcode, physical and abstract meet. The microcode controls the actual circuits.

A stairway of sorts leads down into a computer. At the top stand the high-level languages. A number of these exist, and more are on the way all the time. They vaguely resemble human languages, and they remain the same no matter what computer a programmer uses. Alsing wanted me to write a little program in the high-level language called BASIC, which resembles a pidgin English. In one part of my program I ordered a simple division of two numbers, a command that in BASIC is represented by a slash: "/". I typed the program into Trixie via Alsing's terminal. What happened to the "/"?

Inside Trixie's storage system lay a number of programs called interpreters. There was one for BASIC, and that interpreter program worked on my high-level program. Actually, the interpreter gave Trixie instructions as to how to begin translating my program into instructions that Trixie's circuits could respond to. Computers compute in order to compute. The interpreter for BASIC had Trixie translate the "/" into so-called assembly language.

It is commonplace today for programmers to stick exclusively to high-level languages and never look inside their machines. Alsing felt that they were missing something. He remembered learning assembly language during his time of midnight programming. "It was neat to learn it. I could skip the middleman and talk right to the machine. It was also great for me to learn that priestly language. I could talk to God, just like IBM." Written down outside the machine, assembly language is a list of mnemonics, such as ADD, Skip On Equal, and so on. It contains the names of all the roughly two hundred basic operations that Trixie can perform. Instructions, these operations are called. Inside the machine, these instructions exist, of course, in the form of electrical charges. No single instruction in Trixie's set was equivalent to the "/". The "/" became a series of several discrete instructions.

Not many years before, in most computers, the "/", once it had been translated into its constituent assembly-language instructions, would have been fed right into the actual circuits. A computer in which the stairway ends there, at the level of assembly language, is said to be "hard-wired." It has circuits especially designed to perform each basic operation in the machine's instruction set. But by the seventies — again, in most computers — assembly language no longer went directly to the circuits, but was itself translated into another language, called microcode.

For each assembly-language instruction there exists a microprogram, and most microprograms consist of several microinstructions. Each of Trixie's microinstructions, in turn, consists of 75 bits. Seen written on a page, a microinstruction is a string of 0's and 1's. These correspond directly, of course, to strings of high and low voltages stored in a special place inside the computer — a "microstorage" compartment. Each string of 75 bits is divided into portions, and each portion is destined for some part or parts of the machine's circuitry. The 75 bits of each microinstruction are the actual signals that will make the gates in the circuits open and close in just the right patterns. So my "/" became a linked list of, let's say, 10 assembly-language instructions, each of which became a microprogram, each of which consisted, on the average, of 3 microinstructions, each of which consisted of 75 bits. The simple "/" was now platoons of signals, which were sent out one after the other, causing Trixie's circuits to take the two numbers I had provided for division, to translate these numbers into electrical code, to determine which was to be divided by which, to run the now encoded numbers through the Arithmetic and Logic Unit in such a way as to divide them (a labyrinthine passage itself), and to put the answer somewhere for the next step in my program. Actually, far more microsteps than that occurred. Indeed, the physical machine responds only to microcode. It was microcode, at bottom, that caused Trixie to translate my "/" into microcode. In this sense, the computer chases its tail.

Since I'd asked for it, the result of this one operation in my little program — the quotient of my division — returned through another complex process, at the end of which it appeared on the screen of Alsing's cathode-ray tube in the form of a decimal number. The entire journey, from the moment when I ordered that the program run, to the time when the answer to the division appeared, occurred in an imperceptible moment, no more noticeable than the time it takes for a light to go on after its switch is thrown. But the speed seemed less impressive than the sheer volume of action that had occurred. I had this small revelation: division was, after all, the name of something intricate.

Some modern computers, most notably the machines of Seymour Cray, remain hard-wired; they respond directly to the electrical equivalent of assembly language. Microcode just makes the language more specific. Microcode is, in this sense, like early Old English, in which there was no word for fighting and a poet who wished to convey the idea of battle had to describe one.

The chief advantage of microcode is flexibility, which accrues mainly to the builder of computers. If, for instance, unforeseen defects crop up after a machine has gone to market — and this almost always happens — the manufacturer can often repair them without changing printed-circuit boards. Often the microcode can be altered instead, at considerably smaller cost. Eagle would be designed to make such changes especially cheap and painless. The code would exist on a so-called floppy disk, like a 45-rpm record, not in unalterable ROMs inside the machine. Each morning, when starting up the computer, the operator would play the disk into the circuits of Eagle's microstorage compartment. If the code had to be changed, the engineers could merely change the floppy disk and send a new copy to customers.

Writing microcode, however, is no simple task. The code is by definition intricate. To make the machine execute just one of its two hundred or three hundred basic instructions, the coder usually has to plan the passage of hundreds of signals through hundreds of gates. Limited storage space forces the coder to economize — to make one microinstruction accomplish more than one task, for example. At the same time, though, the coder must take care that one microinstruction does not foul up the performance of another.

Alsing had written fat volumes of microcode, in an extremely quirky way. The Eclipse was to be Data General's first micro-coded machine. Alsing signed up for the job — there was signing-up in those days, too — and then he procrastinated. Month after month, his boss would ask him how the code was coming along and he would say: "Fine. A few problems, but pretty well." In fact, though, he had not yet written a single line of code. Finally, he could sense that his boss and some of his colleagues were growing angry; failure had become an almost palpable object — a pair of headlights coming toward him down the wrong side of a road. Scared, he packed up the necessary circuit diagrams, specs and manuals and went to the Boston Public Library.

The Eclipse contained 195 assembly-language instructions, which in the end Alsing encoded in some 390 microinstructions, many of which performed multiple duties. He said he wrote most of those microinstructions in two weeks at the library. Perhaps it really took him less; West believed that Alsing did it all in two days and nights. "Without question he did," West insisted.

It was always this way with Alsing. The summer before the Eagle project began, he was assigned to write the code for a new Eclipse. As usual, he stalled, and when he felt that he was about to get in trouble, he went home with an armful of books.

He lived about fifteen miles from Data General, in a new ranch house. "My microporch," he said, showing me into a small screened porch. We looked out on a nearby grove of white pines and smelled the needles on the floor of the woods. The room contained a bright green outdoor carpet, an electric grill for barbecues, some uncomfortable-looking wrought iron chairs and a table with a glass top. That summer, he had asked his wife to keep his three sons away from there for a while, had deposited his manuals on the tabletop, and had started to think. Again, he did the entire job in a rush, and finished in about two weeks. "A quick hit," he said.

Much of the engineering of computers takes place in silence, while engineers pace in hallways or sit alone and gaze at blank pages. Alsing favored the porch and staring out at trees. When writing code, he said, he often felt that he was playing an intense game of chess with a worthy opponent. He went on:

"Writing microcode is like nothing else in my life. For days there's nothing coming out. The empty yellow pad sits there in front of me, reminding me of my inadequacy. Finally, it starts to come. I feel good. That feeds it, and finally I get into a mental state where I'm a microcode-writing machine. It's like being in Adventure. Adventure's a completely bogus world, but when you're there, you're there.

"You have to understand the problem thoroughly and you have to have thought of all the myriad ways in which you can put your microverbs together. You have a hundred L-shaped blocks to build a building. You take all the pieces, put them together, pull them apart, put them together. After a while, you're like a kid on a jungle gym. There are all these constructs in your mind and you can swing from one to the other with ease.

"I've done this in short intervals for a short period each year. There's low intensity before it and a letdown at the end. There's a big section where you come down off it, and sometimes you do it awkwardly and feel a little strange, wobbly and tired, and you want to say to your friends, 'Hey, I'm back.' "

If simplicity is elegance, then Alsing's code for the Eclipse was a kludge. It was full of tricks and subtleties — but for a purpose: he had at his disposal a severely limited storage space. And his concoction worked well, in the end. It takes no imagination to see, however, that for someone trying to manage the invention of a computer, Alsing's style could provoke nightmares. Afterward, when the first Eclipse was sent to market, Alsing felt proud. "I did a damn good job," he told himself. He received what he still felt was a touch of glory. He was asked to attend the conference at which the company's salesmen were briefed about the new machine. Grinning, Alsing remembered one of the executives saying to the salesmen, "Okay, here's your brochures. You can jack off to them on the plane home." Alsing was asked to say a few words. The salesmen applauded. "That was pretty rich stuff to me, coming out of the cellar and receiving all that fame and attention." But his habit of putting off the creation of his code until all was almost lost probably did hurt him. During his months of procrastination, he wondered why he couldn't get down to work and came to believe that he was simply lazy, and he guessed that he let it show in little ways — an averting of the eyes, a slump in the shoulders. And partly because of that, perhaps, and certainly because of the anxiety he created, he did not receive a substantial gift of stock for his work on the Eclipse, although some other engineers got tidy rewards.

One time West happened to mention Alsing's name to an executive upstairs, and the executive said, "You know, it's funny, but when you think about it, Alsing's written just about every line of microcode that's come out of Data General." Alsing did make himself easy to overlook. By his own account, he deliberately obscured himself in West's shadow. He went to work for West, he said, after West took over the Eclipse Group, largely because he felt that it would be safer to stand on West's side than not. West, almost overnight, had become formidable. For Alsing, West represented a buffer and a shield. He had some other reasons, though, for signing up to help West on the Eagle project.

Kludge made Alsing imagine a wheel built out of bricks, with wooden wedges in between them; such a thing might work, but no sane engineer would be proud to have designed it. Alsing tended to agree with those who maintained that Eagle must, by definition, be something of a kludge. He did believe it when West said that Eagle could put a lot of money on Data General's bottom line, but past experience made Alsing doubt that he would ever see any of the loot. He even thought sometimes that if he went on helping to pound out 16-bit Eclipses and never worked on another big new machine, he would be a little bored but probably content. Nevertheless, Alsing stood with West from the start of Eagle, and he never asked for coaxing.

"West's not a technical genius. He's perfect for making it all work. He's gotta move forward. He doesn't put off the tough problem, the way I do. He's fearless, he's a great politician, he's arbitrary, sometimes he's ruthless." Alsing laced his hands together so that his knuckles made an arch, on which he rested his chin. Why was West so obsessed about this machine? How would it all end? Sometimes, Alsing said, he felt he had joined up mainly to find out. He went on: "I screamed and hollered over NAND gates and microinstructions with the first Eclipse, but I'm too old to feel that way about computers now. This would be crashingly dull if I was doing it for someone else. West is interesting. He's the main reason why I do what I do."

Looking around the basement, some of the team's brand-new engineers would sometimes wonder what would happen to them when they turned thirty. Being young, they could make light of the question, and say, as one did, "When a computer engineer gets old, he gets turned out to pasture or else made into dog food." Data General was a young company, and so its engineers tended to be young. There really was such a thing in the world as a practicing middle-aged computer engineer. It did appear, however — management handbooks say so — that many engineers experience a change of life when they reach the age of thirty or so.

Among engineers generally, the most common form of ambition — the one made most socially acceptable — has been the desire to become a manager. If you don't become one by a certain age, then in the eyes of many of your peers you become a failure. Among computer engineers, I think, the wish to manage must be a virtual instinct. The industry's short product cycles lend to many projects an atmosphere of crisis, so that computer engineering, which is arduous enough in itself, often becomes intense. The hours are long. Emotions get taxed. Moreover, the technology of computers changes constantly; every year it's a struggle to keep up with the youngsters fresh out of school. What another of West's old hands called "a long-term tiredness" can easily creep over computer engineers in their thirties.

From the start of Eagle, Alsing disengaged himself from much of the technical work on the machine. He was running the Microteam, but from a little distance. Eagle would contain more code than any Data General machine before it — as much code as Alsing had written in his entire career. Alsing could not write all of it, even if there were time. He simply could not generate the excitement he used to feel about gates and bits. Moreover, he believed that since he could not write all of the code, then he couldn't write any of it. These new kids, he saw, approached the job in a way he never had. They worked steadily, day after day, night after night. That was fortunate, for the sake of the team. Alsing admired their discipline. He believed that it exceeded his by far. So he left the writing of the code to half a dozen new recruits, and most of the supervision of their work to submanagers.

Sometimes Alsing worried about his detachment. "Although I sometimes say I don't care too much this time around, if I were to lose this — if I were to be fired or transferred to another project more mundane — I would be, uh, very unhappy. Maybe I'm starting to take this place for granted," he said once.

For a time, when he was still in college, Alsing had wanted to become a psychologist. He adopted that sort of role now. Although he did keep track of his team's technical progress, he acted most visibly as the social director of the Microteam, and often of the entire Eclipse Group. Fairly early in the project, Chuck Holland had complained, "Alsing's hard to be a manager for, because he goes around you a lot and tells your people to do something else." But Holland also conceded: "The good thing about him is that you can go and talk to him. He's more of a regular guy than most managers."

Alsing created the Microteam. He chose its members and he gave them their first training, with some help from Rosemarie Seale. Nowadays it takes a computer to build a new computer, especially when it comes to writing microcode for one. Alsing figured that before the Microkids did anything else, they must learn how to manipulate Trixie. He didn't want simply to give them a stack of manuals and say, "Figure it out." So he made up a game. As the Microkids arrived, in ones and twos, during the summer of 1978, he told each of them to figure how to write a certain kind of program in Trixie's assembly language. This program must fetch and print out the contents of a certain file, stored inside the computer. "So they learned the way around the system and they were very pleased," said Alsing. "But when they came to the file finally, they found that access to it was denied them."

The file in question lay open only to people endowed with what were called "superuser privileges." Alsing had expected the recruits to learn how to find this file and, in the process, to master the system. He was equally interested in seeing what they would do when they found they couldn't get the file.

One after the other, they came to him and said, "I almost have it."

"Okay," said Alsing, "but you don't have it."

In the end, most Microkids went to Rosemarie. Alsing had conferred with her beforehand. She was to help the Microkids find the file, if they asked. They learned something, Alsing felt. "If a person knows how to get the right secretary, he can get everything. It was a resourceful solution — one of the solutions I hoped they'd find."

This first game led to others. Not long after the recruits arrived, the "Tube Wars" began. As a rule, it was the kids against Alsing. In one commonly used gambit, a Microkid would sit down at a terminal and order Trixie to open up Alsing's files. The Microkid would then move the files to a new location. Returning from coffee or lunch, Alsing would find his files gone. He'd hear tittering from the cubicles nearby. And he would know he'd been "tube-warred."

"What did you do to me?" he'd cry.

"Find out, stupid," a voice would answer.

The Microkids weren't the only ones playing games with the computers in the basement. A young woman worked for Rosemarie. She was unmarried and, by general consensus, good-looking. Every day for a couple of weeks during the Eagle project, she was "assaulted" at her desk. She would be doing her electronic paperwork when suddenly everything would go haywire, all her labor would be spoiled, and on the screen of her cathode-ray tube would appear cold, lascivious suggestions. "Whoever was doing it," said West, had "the mentality of an assassin."

West put Alsing on the case. Alsing had some members of the team lay traps inside the computer system — traps designed to leave a trail back to the masher's terminal. But the masher spotted all of these; one time he made his escape by bringing to an abrupt halt the entire system on which most of the engineering departments relied. He had to be stopped, and eventually Alsing found a strong suspect, a young man outside the group. Alsing had a casual chat with him about all the marvelous tricks that could be played with the in-house computing system; afterward, the obscene messages ceased. Wholesomeness, in this regard, returned to the basement. Indeed, said one young engineer, the place seemed antiseptic.

The masher's game had been especially nasty and unfair, Alsing pointed out, because the victim could not fight back. But Tube Wars pitted worthy adversaries against each other. The jousting did no harm, and, on the contrary, released tension. One day Alsing came back from lunch and went to work at his terminal. Everything looked right, all his files seemed to be in place — until he tried to do something with them. Then, to his surprise, he found that all of them were vacant. "It was like opening a filing cabinet and finding all the folders empty. They were dummy files. It took me an hour to find the real ones. So now I can never be sure, when I log on the system, that what I see is real."

Alsing struck back. He created an encrypted file and tantalized the team, "There's erotic writing in there and if you can find it, you can read it." They tried, and ultimately all gave up, including Bob Beauchamp. Alsing taunted Beauchamp, though. So Beauchamp tried again. This time he wrote a program that broke Alsing's encryption system. "He beat me," Alsing said. "But I think he was too much of a gentleman to read what was inside."

Alsing double-encrypted the secrets in his files after that, and for many months he assumed they were safe. Beauchamp abandoned his first approach, feeling that it was a little crude. Now he made a slight revision in Trixie's operating system. In essence, he instructed the machine that whenever Alsing encrypted a message, the operating system should send to Beauchamp's files an unencrypted copy. This was the ultimate victory in Tube Wars, not least of all because Alsing never learned what Beauchamp had done to him until Beauchamp himself spilled the beans.

Tube Wars died out slowly. At their height, whenever I visited Alsing, I'd take a look at the screen of his cathode-ray tube and almost always see something peculiar written on it, some message or picture sent to him by a young engineer at play. I'd come into Alsing's cubicle and there on the screen would be a picture of a fist with the middle finger extended; or there'd be a little story on his screen:

SEX LIFE OF AN ELECTRICIAN (PART 3) FULLY EXCITED MILLIE AMP MUMBLED OHM! OHM! OHM! . . .

Alsing arranged several sorts of social gatherings, among them a weekly meeting of the Microteam held around a table in a barren little conference room that contained one tiny window. I went to a couple of these convocations. Alsing would call the assembly to order, read a few announcements, and then submit to teasing.

"I read about someone who did a study of his company and discovered that he was the least important employee. So he quit," Alsing said.

"So long, Carl!" cried one of the Microkids.

There was a lot of talk about Alsing's idea that they should hand out Honorable Member of the Microcode Group Awards.

"I think we oughta give one to West," said one of the team. "So that when we get pissed off at him, we can take it back."

"No," said another, "because now he'll be a member of the Microteam. He can solve his own problems."

They did do some business, using the beautiful and, to me, in-scrutable language of the microelectronic era: *hexaddresses, de-fault redix, floating-point mantissas, swapbites, sys log, sim dot, scratch pad.*

"The scratch pad doesn't come alive on CPD until one-sixteen," said one.

"That means all the stack tests go away!" cried another.

"That's right."

There was laughter all around the table.

When it died away, a Microkid said, "Look, we can speed up Eagle's stack stuff by putting in a scratch pad."

"Can we," asked another, "also plant a little bomb in the thing?"

Alsing sat with his hands folded, smiling subtly. He looked like a big contented cat. No one actually called an end to the meeting. It simply petered out in laughter.

And yet there was bad feeling among some of them, much of it directed toward Alsing; and even then, petty intrigue was in progress. Some of the team would eventually describe the weekly micromeetings as "Alsing's weekly No-Op" — No-Op being the name of an assembly-language instruction that accomplishes nothing. For a long time, however, almost all of the recruits enjoyed these meetings, the Tube Wars and other entertainments that Alsing arranged. He made a point of sharing lunchtime with some of them several days a week. And they appreciated Alsing's friendliness; they could always talk to him.

Alsing believed that the team's managers, in handling the new recruits, really were practicing what was called "the mushroom theory of management." It was an old expression, used in many other corners of corporate America. The Eclipse Group's managers defined it as follows: "Put 'em in the dark, feed 'em shit, and watch 'em grow." It was a joke with substance, Alsing felt; and he believed that their mushroom management needed an occasional antidote. Alsing in effect had signed up to provide the kids with some relief from their toil. West warned him several times, "If you get too close to the people who work for you, Alsing, you're gonna get burned." But West didn't interfere, and he soon stopped issuing warnings.

One evening, while alone with West in West's office, Alsing said: "Tom, the kids think you're an ogre. You don't even say hello to them."

West smiled and replied, "You're doing fine, Alsing."

6

FLYING UPSIDE DOWN

WEST OFTEN SAID that they were playing a game, called getting a machine out the door of Data General with their names on it. What were the rules?

"There's a thing you learn at Data General, if you work here for any period of time," said West's lieutenant of hardware, Ed Rasala. "That nothing ever happens unless you push it." To at least some people upstairs, this condition took the name "competition for resources." As a strategy of management, it has a long lineage. *"Throw down a challenge,"* writes Dale Carnegie in that venerable bible of stratagems dressed up as homilies, *How to Win Friends and Influence People.*

In a sense, the competition between Eagle and North Carolina was institutionalized; each project lay in the domain of a different vice president. But that may have been accidental. West's boss, who was the vice president of engineering, Carl Carman, remarked that he had worked at IBM and that compared to competition among divisions there, rivalry among engineering teams at Data General resembled "Sunday school." Moreover, Carman said, in a company with a "mature product line" like Data General's, situations naturally occur in which not enough large new computers are needed for every team of computer builders to put one of its own out the door. "And yeah," Carman continued, "the competition is fostered." He said that de Castro liked to see a little competition stirred up among teams. Let them compete with their ideas for new products, and bad ideas, as well as the negative points of good ones, are likely to get identified inside the company and not out in the marketplace. That was the general strategy, Carman said. What it now meant downstairs, to the Eclipse Group, was that they not only had to invent their new computer but also had to struggle for the resources to build it. Resources meant, among other things, the active cooperation of such so-called support groups as Software. You had to persuade such groups that your idea had merit and would get out the door, or else you wouldn't get much help — and then your machine almost certainly *wouldn't* get out the door.

Here's how it looked to West: The company could not afford to field two new big computers; Data General had made a large investment in North Carolina as a place where major computers would be built; and although the Eclipse Group's engineers had good technical reputations, North Carolina's had better ones. The game was fixed for North Carolina and all the support groups knew it.

So West started out by calling Eagle "insurance" — it would be there in case something went wrong down south. Thus he avoided an open fight and thus he could argue that the support groups should hedge their bets and put at least a little effort into this project, too. As for North Carolina's superior reputation, West never stopped suggesting to people around Westborough that their talents had been slighted. His message was: "Let's show 'em what we can do."

"West takes lemons and makes lemonade," observed Alsing.

From the first rule — that you must compete for resources — it followed that if your group was vying with another for the right to get a new machine out the door, then you had to promise to finish yours sooner, or at least just as soon as the other team promised. West had said that the Eclipse Group would do EGO in a year. North Carolina had said, okay, they'd finish their machine in a year. In turn, West had said that Eagle would come to life in a year. West said he felt he had to pursue "what's-the-earliest-date-by-which-you-can't-prove-you-won't-be-finished" scheduling in this case. "We have to do it in a year to have any chance." But you felt obliged to set such a schedule anyway, in order to demonstrate to the ultimate bosses strong determination.

Promising to achieve a nearly impossible schedule was a way of signing up — the subject of the third rule, as I saw it. Signing up required, of course, that you fervently desire the right to build your machine and that you do whatever was necessary for success, including putting in lots of overtime, for no extra pay.

The fourth rule seemed to say that if the team succeeded, those who had signed up would get a reward. Not one in the group felt certain that stock options were promised in case of success. "But it sure as hell was suggested!" said one of the Microkids. All members of the team insisted that with or without the lure of gold, they would have worked hard. But for a while, at least, the implied promise did boost spirits, which were generally high anyway.

I think that those were the rules that they were playing by, and when I recited them to some of the team's managers, they seemed to think so, too. But Alsing said there was probably another rule that stated, "One never explicitly plays by these rules." And West remarked that there was no telling which rules might be real, because only de Castro made the rules that counted, and de Castro was once quoted as saying, "Well, I guess the only good strategy is one that no one else understands."

They lived in a land of mists and mirrors. Mushroom management seemed to be practiced at all levels in their team. Or perhaps it was a version of Steve Wallach's ring protection system made flesh: West feeling uncertain about the team's real status upstairs; West's own managers never completely aware of all that their boss was up to; and the brand-new engineers kept almost completely ignorant of the real stakes, the politics, the intentions that lay behind what they were doing. But they proceeded headlong. Wallach's architectural specification was coming along nicely now.

The attempt to turn those ideas into silicon and wire and micro-code had begun. Now they had to create a complete design and do it in a hurry. Carman made it policy that members of the team could come and go more or less as they pleased. These were confident, aggressive young engineers — "racehorses," West liked to say — and they were about to be put under extreme pressure. Carman hoped that by allowing them to stomp out of the basement at any time without fear of reprisal, he would be providing an adequate "escape valve."

At last, by the fall of 1978, the preliminaries were complete. The kids had been hired, the general sign-up had been performed, the promises suggested, and the escape valve established. Then West turned up the steam.

You're a Microkid, like Jon Blau. You arrived that summer and now you've learned how to handle Trixie. Your immediate boss, Chuck Holland, has given you a good overall picture of the microcode to be written, and he's broken down the total job into several smaller ones and has offered you your choice. You've decided that you want to write the code for many of the arithmetic operations in Eagle's instruction set. You always liked math and feel that this will help you understand it in new, insightful ways. You've started working on your piece of the puzzle. You can see that it's a big job, but you know you can do it. Right now you're doing a lot of reading, to prepare yourself. Then one day you're sitting at your desk studying Booth's algorithm, a really nifty procedure for doing multiplication, when Alsing comes by and tells you, "There's a meeting."

You troop into a conference room with most of the other new hirees, joking, feeling a little nervous, and there waiting for you are the brass: the vice president of engineering, another lower-level but important executive, and West, sitting in a corner chewing on a toothpick. The speeches are brief. Listening intently, you hear all about the history of 32-bit superminis. These have been around awhile, but sales are really picking up. DEC's starting to turn out VAXes like jelly beans, and the word is DEC'll probably introduce a new model of VAX in about nine months. No one's saying it's your fault, but Eagle's late, very late. It really must be designed and brought to life and be ready to go by April. Really. In just six months. That won't be easy, but the brass think you can do it. That's why you were hired — you're the cream of a very fine crop. Everything depends on you now, they say.

You feel good about yourself and what you're doing when you leave that meeting. You go right back to your desk, of course, and pick up Booth's algorithm. In a little while, though, you feel you need a break. You look around for another Microkid to share coffee with you. But everyone is working, assiduously, peering into manuals and cathode-ray tubes. You go back to your reading. Then suddenly, you feel it, like a little trickle of sweat down your back. "I've gotta hurry," you say to yourself. "I've gotta get this reading done and write my code. This is just one little detail. There's a hundred of these. I better get this little piece of code done today."

Practically the next time you look up, it's midnight, but you've done what you set out to do. You leave the basement thinking: "This is life. Accomplishment. Challenges. I'm in control of a crucial part of this big machine." You look back from your car at the blank, brick, monolithic back of Building 14A/B and say to yourself, "What a great place to work." Tomorrow you'll have to get to work on an instruction called FFAS. That shouldn't be too hard. When you wake up the next morning, however, FFAS is upon you. "Oh my God! FFAS. They need that code next week. I better hurry."

"The pressure," said Blau. "I felt it from inside of me."

In another cubicle, around this time, Dave Epstein of the Hardy Boys is dreaming up the circuits of a thing called the Microsequencer. Nothing else will work without this piece of hardware.

Some weeks ago, Ed Rasala asked Epstein, "How long will it take you?"

Epstein replied, "About two months."

"Two months?" Rasala said. "Oh, come on."

So Epstein told him, "Okay, six weeks."

Epstein felt as if he were writing his own death warrant. Six weeks didn't look like enough time, so he's been staying here half the night working on the thing, and it's gone faster than he thought it would. This has made him so happy that just a moment ago he went down the hall and told Rasala, "Hey, Ed, I think I'm gonna do it in four weeks."

"Oh, good," Rasala said.

Now, back in his cubicle, Epstein has just realized, "I just signed up to do it in four weeks."

Better hurry, Dave.

"I don't know if I'm complaining, though," says Epstein. "I don't think I am. I work well under pressure." Indeed, Epstein will finish on schedule and his design will turn out to be almost errorless.

But not everyone works well under such conditions. Not everyone thinks it is worth it. A couple of engineers have already dropped out. A few are less than happy. One Hardy Boy, Josh Rosen, looks around and can hardly believe what he sees. For example, Microkids and Hardy Boys are arguing. A Microkid wants the hardware to perform a certain function. A Hardy Boy tells him, "No way — I already did my design for microcode to do that." They make a deal: "I'll encode this for you, if you'll do this other function in hardware." "All right."

What a way to design a computer! "There's no grand design," thinks Rosen. "People are just reaching out in the dark, touching hands." Rosen is having some problems with his own piece of the design. He knows he can solve them, if he's just given the time. But the managers keep saying, "There's no time." Okay. Sure. It's a rush job. But this is ridiculous. No one seems to be in control; nothing's ever explained. Foul up, however, and the managers come at you from all sides.

"The whole management structure," said Rosen. "Anyone in Harvard Business School would have barfed."

In relatively serene times, some years before Eagle, West and his wife had made friends with an electrician who lived in their town. The man's name was Bernie. He owned a small airplane. Since West's farmhouse and barns lay under one line of approach to the little local airport, Bernie often flew over. When he did, he would waggle his wings, he might do a quick roll; sometimes he'd climb halfway out of his window and wave down at the Wests. "Bernie likes to fly upside down," West remarked, and he and his wife shook their heads and laughed.

Alsing often heard West talk about flying upside down. It seemed to mean taking large risks, and the ways in which West used the phrase left Alsing in no doubt that flying upside down was supposed to be a desirable activity — the very stuff of a vigorous life.

Ed Rasala allowed that West made life in their corner of the basement more dramatic — "definitely more dramatic" — than it usually had to be. But neither Rasala nor Alsing nor Wallach balked when West said that they had to fly upside down now. Over this project loomed the memory of EGO. No one wants to see hard work come to nothing, and EGO was generally accounted a disaster. But it had lasted just a few months and had involved only a few engineers. About thirty were working on Eagle now. That the project might be tossed on the scrap heap somewhere along the way, after months of thirty souls' passionate labor, was unthinkable. But it could certainly happen, they thought. West had felt that he had to promise to do Eagle in something like a year in order to get the chance to do it. Now he chose to believe that to get it out the door, they really had to come close to meeting that absurd schedule. At the same time, they had to do it right — right in the commercial sense. The whole project was risky, from the start. In service of the big risk, West undertook on the team's behalf many smaller ones.

"We're always assuming that things'll break right for us," observed Alsing. West was assuming, for instance, that the software resources would be there when they were needed. They were all assuming that youngsters fresh from college could build a major new computer, though none of the recruits had built anything like this before.

Looking for a technical advantage, West gambled that the coming thing in chips was a type of circuit known as a PAL. The manufacture of integrated circuits is a fairly risky business; it is said that factories can suddenly become inoperative for no apparent reason — though a small infusion of dust is a common suspect. So the conventional wisdom holds that in making a new computer, you never plan on using any sort of brand-new chip unless at least two companies are making it. At that moment, only one fairly small company was making PALs. But if PALs really were the coming thing, it would be a win to use them. West decided to do so.

West figured that the Eclipse Group had to show quick and constant progress in order to get the various arms of the company increasingly interested in helping out. For public relations, and maybe in order to keep the pressure on his crew, he made extravagant claims. He always pushed them one step ahead of themselves. Before Wallach finished specifying the architecture, West had the team designing the boards that would implement the architecture; before the engineers cleaned up their designs, West was ordering wire-wrapped, prototype boards; before the wirewraps could possibly be made right, he was arranging for the making of printed-circuit boards; and long before anyone could know whether Eagle would become a functioning computer, West had the designers stand in front of a TV camera and describe their parts of the machine. The result of this last act of hubris was a videotaped extravaganza some twenty hours long. West planned to use it, when the right time came (if it ever did), as a tool for spreading the news of Eagle all around Westborough. "Pretty gutsy," he said, with a grin, nodding toward the shelfful of video casettes.

One evening West paused to say to me: "I'm flat out by definition. I'm a mess. It's terrible." A pause. "It's a lot of fun."

West established the rules for the design of Eagle and he made them stick. The team should use as little silicon as possible, a mere few thousand dollars' worth of chips. The CPU should fit on far fewer than VAX's twenty-seven boards, and each major element of the CPU should fit on a single board. If they could fulfill those requirements, Eagle would be cheaper to build than VAX. On the other hand, it had better run faster than VAX, by certain widely accepted standards. It should be capable of handling a host of terminals. A CPU is not a functioning computer system; Eagle also had to be compatible with existing lines of Data General peripherals as well as with Eclipse software.

On the Magic Marker board in his office, West wrote the following:

Not Everything Worth Doing Is Worth Doing Well.

Asked for a translation, he smiled and said, "If you can do a quick-and-dirty job and it works, do it." Worry, in other words, about how Eagle will look to a prospective buyer; make it an inexpensive but powerful machine and don't worry what it'll look like to the technology bigots when they peek inside. West espoused these principles of computer design: "There's a whole lot of things you've gotta do to make a successful product. The technological challenge is one thing, but you can win there and still have a disaster. You gotta give 'em guidelines so that if they follow them, they're gonna be a success. 'Do ABC and D without getting the color of the front bezel mixed up in it.' " Another precept was "No bells and whistles." And a third: "You tell a guy to do this and fit it all on one board, and I don't want to hear from him until he knows how to do it."

West reviewed all of the designs. Sometimes he slashed out features that the designers felt were useful and nice. He seemed consistently to underestimate the subtlety of what they were trying to do. All that a junior designer was likely to hear from him was "It's right," "It's wrong," or "No, there isn't time."

To some the design reviews seemed harsh and arbitrary and often technically shortsighted. Later on, though, one Hardy Boy would concede that the managers had probably known something he hadn't yet learned: that there's no such thing as a perfect design. Most experienced computer engineers I talked to agreed that absorbing this simple lesson constitutes the first step in learning how to get machines out the door. Often, they said, it is the most talented engineers who have the hardest time learning when to stop striving for perfection. West was the voice from the cave, supplying that information: "Okay. It's right. Ship it."

West kept final authority over the circuit designs. But he loosened control over most of the management of their creation. How did the Hardy Boys invent the general plan for the hardware? "Essentially," said Ed Rasala, "some of the guys and I sat down and decided what elements we needed." Over in the Microteam, though never explicitly told to do so, Chuck Holland took on the job of organizing the microcoding job. Holland and Ken Holberger mediated the deals between Hardy Boys and Microkids, but in general the veterans let them work things out for themselves. The entire Eclipse Group, especially its managers, seemed to be operating on instinct. Only the simplest visible arrangements existed among them. They kept no charts and graphs or organizational tables that meant anything. But those webs of voluntary, mutual responsibility, the product of many signings-up, held them together. Of course, to a recruit it might look chaotic. Of course, someone who believed that a computer ought to be designed with long thought and a great deal of preliminary testing, and who favored rigid control, might have felt ill at the spectacle. Criticism of that sort flattered West. "Show me what I'm doing wrong," he'd say with a little smile.

In fact, the team designed the computer in something like six months, and may have set a record for speed. The task was quite complex.

The machine took its first material form in paper — in a fat volume of pages filled up with line after line of 0's and 1's, and in bound books large as atlases that contained the intricate geometrical depictions of the circuits, neatly drawn by their draftsman. You could think of this small library of microcode and schematics as the engineers' collected but not wholly refined thoughts on a variety of subjects. The language was esoteric, but many of the subjects were as familiar as multiplication.

I had imagined that computer engineering resembled the household electrician's work, but it seemed the bulk of it lay in making long skeins of logical connnections, and it had little to do, at least at this stage, with electricity. I wondered, too, why they had to struggle to fit Eagle's CPU onto seven boards — seven was the goal — when elsewhere engineers were routinely packing entire CPUs onto single chips. The general answer was that a multi-board CPU performs simultaneously many operations that a single-chip CPU can do only sequentially. By making a CPU on several boards, you can make it run much faster than a CPU on a chip. A time was probably coming when components would operate so quickly that the distance that signals had to travel would intimately affect the speed of most commercial computers. Then miniaturization and speed would become more nearly synonymous. But that day had not yet arrived.

Designing Eagle, they went far beyond what could yet be done on a single chip. Many of the chips that they used came to them ready-made to perform certain kinds of operations. This took most of the pure physics from their endeavor. West's decision to use the new chips called PALs gave the designers certain advantages. Certainly it helped some of them work swiftly. When Ken Holberger, for instance, came to a tricky part of a design, one that promised to take a long time to create, he could often just draw a box on his diagram and leave that box blank. A single programmable PAL could be made to perform all the functions that the box had to do. "PAL here," he would write on his diagram, in effect. Later, he could come back and program the PAL. In the meantime, prototypes of the new machine could be built. But eventually the designers had to program every PAL. They had to invent the complex internal organization of each of those chips. They had to know what all the chips were meant to do. And there were thousands of chips in an Eagle.

Some engineers likened the chips to an unassembled collection of children's building blocks. Some referred to the entire realm of chip design and manufacture as "technology," as if to say that putting the chips together to make a computer was something else. A farmer might feel this way: "technology" is the new hybrid seeds that come to the farm on the railroad, but growing those seeds is a different activity — it's just raising food.

Eagle's designers adopted ideas that were abroad in the industry. From their files they took some of the material that had been concocted in the days of EGO and Victor; much of the inspiration for the details came from EGO. But they invented the largest part of Eagle themselves, in five or six months, between the late summer of 1978 and the new year.

From one angle, the task was to make an engine that would obey — without fail and at great speed — each of the roughly four hundred chores named in Eagle's instruction set. One night after work, Microkid Jon Blau described for me the progress of just one of these basic chores, or assembly-language instructions, through the engine. He left out some steps, simplified many others, and kept the description fairly abstract. Even so, the short tour consumed several hours.

Blau was living in an apartment complex near Natick, Massachusetts. His rooms were neat but sparsely furnished, with an old couch, a beanbag-type chair, and a bookcase, eclectically stocked — his intellectual diet seemed to consist of science fiction, well-reviewed novels, philosophy, physics and math. A ten-speed bike leaned against a wall in the bedroom. His bed was a mattress on the floor. He apologized for the emptiness of the refrigerator. It was the apartment of a young man going elsewhere, and it made me feel old.

With Blau, I descended, as it were, into Eagle's engine room. He pointed out the main parts of Eagle and spoke of them as if they were sensate things that asked questions, looked up answers, sent and received messages. It's a way of talking about computers that makes some people nervous, but it was one of the ways he thought about the machine when he was working on it.

In order to execute an instruction in a user program, the engine has to do a lot of other work first, Blau explained. We should assume that a special program — a program of programs, let's call it — is already running. This program communicates with people using the computer and it does jobs for them. It schedules and monitors all the user programs that are currently running. It will also find a program for a user. This program of programs contains input and output assembly-language instructions, which tell the Input/Output Control board — the IOC — how to move information back and forth between the engine and the users' terminals. (The IOC makes it possible for the computer to communicate with the outside world; roughly speaking, it plays the role of a translator who knows many different tongues. It can arrange for the movement of information at high speed and also for communication with user devices that work at relatively slow speeds. It's complicated.)

Okay, said Blau, someone out there at a terminal, a user, wants to run a program; call it "program FOOBAR." Through a terminal, this user tells the program of programs: "Run program FOO-BAR." The program of programs tells the IOC to move *part* of that program from a storage disk outside of the CPU into the CPU's "Main Memory," and to do so at high speed. This accomplished, the program of programs turns over control to program FOOBAR. The machine then starts executing the instructions of program FOOBAR, one after the other.

Before the engine can execute an instruction, it has to find it, of course. Then it has to fetch it and decode it. Thus we came to the printed-circuit board known as the Instruction Processor — the IP, it was called, and it was quite a tricky thing. The IP has a relatively small memory of its own. In a sense, the IP makes assumptions about what the next instructions in the user program will be, and it keeps those instructions handy in its storage. Acting on its assumptions, it finds, fetches, and decodes instructions ahead of time. It "gets them in the pipeline." For this reason, the IP is also known as an accelerator; it does work that any computer must do, but it does it in advance.

Suppose, however, said Blau, that program FOOBAR has been running for a while and the IP discovers, as it does now and then, that it doesn't have the next instruction in its own memory. At this point the IP sends a message to the Address Translation Unit — the ATU — which among other roles, keeps a map of the CPU's Main Memory. In its message, the IP asks the ATU for the location of the next block of instructions in user program FOOBAR.

Let's assume, said Blau, that the ATU finds that next block in its map. This means that the needed instructions are residing in Main Memory. The ATU knows where and passes the news to the IP. The IP, in turn, sends a message to another accelerator, known as the System Cache — Sys Cache, for short — asking for those next instructions.

Maybe the Sys Cache has that next block of instructions. If so, it doesn't have to send away for them, but can pass them right on to the IP, and time is saved. If, however, the Sys Cache doesn't find those instructions in its own memory, it retrieves them from Main Memory and passes them on to the IP.

"But," said Blau, "this next block of instructions may not be in Main Memory either." Other people are using the computer, running other programs. Main Memory is large but not large enough to hold all the parts of all those programs at the same time. Usually it's holding only some parts of each program. So we should assume, said Blau, that the next section of program FOO-

BAR is sitting outside the CPU, still in peripheral storage, on a disk. In this case, the ATU doesn't find the address of the next block of instructions in its map. So the ATU sends a message to the Microsequencer. (The sequencer holds the microcode, and it was worth noting, Blau said, that many of the actions so far undertaken have been performed at the direction of microcode.) The Microsequencer responds to the ATU's message by sending out a certain microprogram, which turns over control of the search to system software. System software, in turn, orders up something called a "page fault program," which contains many assembly-language instructions and hence many microprograms. This page fault program should tell the machine how to find that next block of instructions in program FOOBAR and how to bring it into the memory system.

It really should not happen — it would constitute a software design error if it did, Blau said — but just imagine that the instructions for the page fault program aren't in the memory system but are instead out on a disk somewhere. In order to get the instructions for the page fault program off the disk, Eagle would have to perform a page fault program, but it couldn't perform a page fault program to get the instructions until it already had them. It would be as if you had locked up a cabinet and left the key inside, Blau said. "If this happens, the machine will fall down and endless series of mirrors."

Such a flaw — flaws are often known as "crocks" — is so notorious as to have become uncommon, but Eagle's designers prepared for it just in case. Via the Console Control board — the CC — Eagle is intimately connected with a microcomputer, a standard Data General microNova, which acts in part as Eagle's therapist. If Eagle goes haywire, the microNova should still function; thus, it can run diagnostic programs to find out what's wrong with the larger machine even if the larger machine is dead. The microNova also constantly paws through the bigger machine, on the lookout for such special deadly crocks as a page fault within a page fault. If it detects one, the CC board will be alerted and it will send back a message to the microNova, which will send a warning to the system console, the big typewriterlike machine that sits next to Eagle's CPU, and the console will print out the following:

INFINITE PAGE FAULT. CPU HALTED.

If you're the operator of the system and you see this message, you put in a panicky call to Data General right away.

But suppose none of that happened, said Blau. Suppose the next block of instructions needed for program FOOBAR is easily found. The Sys Cache sends that block to the IP. The IP throws away another block of information to make room for this one, and now the search is over at last. Eagle can begin to execute the next instruction in program FOOBAR.

The IP examines the encoded instruction and determines that it is a "Skip On Equal" — one of those instructions that tells the computer to compare two values and, depending on the result, to follow one of two paths. In assembly language it's written as "WSEQ" (which could be the sound a rusty gate makes).

Specifically, WSEQ instructs the machine to compare two values and if they are equal, to skip the next assembly-language instruction in program FOOBAR and go on to the instruction after it. Assume, said Blau, that the two values — electrically encoded, of course — have found their way into the Arithmetic and Logic Unit, the ALU (sometimes known as "the number cruncher," the heart of any computer).

The IP has determined from the WSEQ instruction several items of information. The most important of these is the address of the microprogram that will tell Eagle exactly what to do in order to Skip On Equal.

The Microsequencer is asking for the address of the next microprogram. The IP now sends this address to the sequencer and the sequencer starts running the microprogram. The IP has also derived at this time the location of the two values that are going to be compared. These packets are already in the ALU. The IP tells the ALU exactly where they are.

"Now we're moving down to a lower level of abstraction," Blau said. "We're going down to where the microcoder lives."

There's a clock inside Eagle. It ticks every 220 billionths of a second. Between each tick of the clock, Eagle performs one microinstruction.

"Tick," said Blau.

The sequencer sends out the first microinstruction of the WSEQ microprogram. The microinstruction is 75 bits — 75 units of high and low voltages. They scatter throughout the circuits. Some go to the ALU and tell it to subtract one of the two values from the other. Some of these microbits go to the IP, some to the ATU and IOC, and some go right back to the sequencer and tell it the address of the next microinstruction in the WSEQ microprogram.

"Tick."

The sequencer sends out that second microinstruction. Part of it goes to the ALU, telling the ALU to examine the result of its subtraction. If the result of the subtraction is not zero — if, in other words, the two values are not equal — then the ALU sends a low voltage down a certain wire. The microinstruction has told the Microsequencer to monitor this wire. Finding a low voltage there, the sequencer brings the WSEQ microprogram to an end. It sends a message to the IP, saying, in effect, "Not equal, no skip on equal this time — get the next assembly-language instruction in the user program."

But if the result of the ALU's subtraction is zero — if the two values are equal — then the ALU sends a high voltage down the special wire that the Microsequencer is monitoring. The sequencer interprets this and sends out the third and last microinstruction in the WSEQ microprogram.

"Tick."

Out goes this last line of microcode, telling all the boards to wait, except for the IP. The IP is told to skip the next assembly-language instruction in the user program and to resume operations with the next instruction after that. This, finally, is how Eagle skips over one crossroad in a program and starts down another road.

West had said that designing a computer was "a mind game." I asked Blau whether following an instruction around through the engine in the way that we had was the sort of mind game that they played when designing Eagle. "You bet it is!" he said. But they played hundreds of such mind games; and figuring out how to equip Eagle merely to WSEQ was itself much more intricate than Blau could fully explain in a single sitting.

Such games of logic, especially if they are played in a hurry — while flying upside down — can take a grip on an engineer's thoughts and hold on. After playing this way for a while, you look at a tree and, aha, it is clear that a tree is much like a computer; and a road with side streets is — what else? — a kind of computer program. Chuck Holland said that this unpleasant sensation, of being locked inside the machine, usually lingered three days — on the rare occasion when he got away from the basement for that long.

West would sit at his desk and stare for hours at the team's drawings of the hardware, playing his own mind games with the results of the other engineers' mind games. Will this work? How much will this cost? Once, someone brought a crying baby past his door, and afterward it took him an hour to retrace his steps through the circuit design he had been pondering. Laughter outside often had the same effect, and once in a while it made his hands shake with rage — especially if he didn't like the design he'd been staring at.

West usually drove out of Westborough fast after work. "I can't talk about the machine," he said one evening, bent forward over the steering wheel. "I've gotta keep life and computers separate, or else I'm gonna go mad."

7

LA MACHINE

ONE MORNING shortly after the designing, West was sitting in his office with the door closed, staring at schematics.

Deadlines lay impossibly near at hand. One of the group's veterans, observing the scene outside West's office, remarked: "The tension among the kids was phenomenal. I could just feel it." Now, in the cubicle across from West's door, a couple of members of the Microteam began to laugh — first one, then the other, then both of them, more and more raucously.

A moment later, Carl Alsing's phone rang.

It was West. "If you don't shut those guys up I'm gonna kill 'em!"

"He was ripshit," said Alsing. "I had to go out there and tell them not to laugh. It was awful. I felt so embarrassed. I felt like one of those old supervisors from the 1800s who used to hire children and make them work eighteen hours a day."

West took a day off and went to look at sailboats.

One Sunday morning — the team's official day of rest — West was at home, trying to lose himself in the newspapers, when the water pump in his house broke down. Throwing the papers aside, he went to the basement to fix the thing. For him, this was a trivial task. He knew just how to do it. He had scarcely begun, though, when hitting some minor setback, he grabbed the pump and hurled it across the room. He stood there, staring after it. If he had done that, what might he do next? In order not to find out, he went upstairs, told his wife to call the plumber, and went right to bed, at midday.

What was wrong with West? Wasn't the worst behind them, now that they had finished the design? He didn't want to talk about it. He merely shook his head.

Now, early in the new year, something like the computer existed, in the form of two partially assembled prototypes. But Eagle wasn't even the equivalent of a pocket calculator yet. The team now had to make that computer work. They called this part of the project "debugging." West had told Ed Rasala to draw up a debugging schedule that would bring Eagle to life by April, the date West had named for the bosses. West had gotten Manufacturing to send down some senior technicians from the factory in Portsmouth, New Hampshire, both to bolster the ranks of debuggers and to get Manufacturing involved. Rasala had put the Hardy Boys on two shifts. West had prescribed eight hours of work for Saturdays. They had made some progress, but now it was coming painfully and slowly. In the local idioms, they were "moving three steps forward, two steps back," and the debugging schedule was "slipping a week a week."

West had assumed that debugging Eagle would resemble the debugging of Eclipses. He had been deceived. You had to make many parts of Eagle start working, it now appeared, before you could really begin to fix it — this because of certain new features, which everyone agreed were "sexy." West couldn't ignore the evidence; he didn't know how to debug this machine, and he could not make himself believe that Rasala and his Hardy Boys would figure it out for themselves.

That fall West had put a new term in his vocabulary. It was *trust.* "Trust is risk, and risk avoidance is the name of the game in business," West said once, in praise of trust. He would bind his team with mutual trust, he had decided. When a person signed up to do a job for him, he would in turn trust that person to accomplish it; he wouldn't break it down into little pieces and make the task small, easy and dull.

To Alsing, West still had that knack for making the ordinary seem special, and the way West said "Trust" made Alsing wonder whether either of them had ever heard the word before. But West prided himself on his skills at debugging and, by repute, he excelled at it. He wanted to go into the lab and will the machine into life, Alsing thought. But if West barged in there now, he would be admitting that he didn't trust his team after all. So West was staying away from the lab and instead was banning laughter outside his door and throwing water pumps around. Most every day now West called Alsing into his office, closed the door, and asked, "What's really going on in the lab, Alsing?"

Wasn't all this excessive? If you set a preposterous schedule, don't you figure that it can slip a little? True, said Alsing, but not to the point. "If you say you're gonna do it in a year and you don't take it seriously, then it'll take three years. The game of crazy scheduling is in the category of games that you play on yourself, in order to get yourself to move."

It was a game in which new hands were always being dealt, a little like poker perhaps. West and his staff had created the deadline of April and, in the act, had agreed at least to pretend to take it seriously. Many months later, Carl Carman would say that no one upstairs believed they would finish Eagle that soon. Some evenings downstairs, West seemed to say the same thing. "We're gonna finish this sucker by April, Alsing," he'd say.

"Yeah, Tom. Sure we are," Alsing would reply.

They'd smile at each other.

Sometimes, however, when, for instance, Alsing came in and told West that the Microteam would probably miss some intermediate deadline, West would say, "Come on, Alsing, this schedule's real."

Now, on an evening early in 1979, Carman brought West a piece of momentous news: North Carolina was going to miss their own deadline by a huge margin.

What did this mean? That the game of internal competition had essentially ended and the game of outlandish scheduling had begun in earnest? West had always maintained that Eagle was crucial to the company. Well, events were proving him right. But that was no cause for celebration. Meeting that preposterous deadline of April, when Eagle should be free of all bugs, was no longer merely desirable or just a matter of pride: it had become a corporate necessity.

At least that was the message West thought Carman gave him — the message West decided to receive, anyway. West sat in his office brooding about it for a long time.

Depending on who you were, you saw West's narrow office either as the lair of an ogre or as a haven with a door behind which you could talk privately. And if it meant the latter, it was, depending on the day of the week and the time of the day, either a place to talk nervously about impending disasters or else to use as a kind of lounge. Steve Wallach, Carl Alsing and Ed Rasala looked forward to their collective weekly meetings with West. He held them in his office on Fridays at 3:00 P.M. They'd do some business, West would tell them the latest company gossip, and, like Alsing at his own meetings with subordinates, West might submit to what Alsing called "zingers from his troops." "We could be in a lot of trouble here," West might say, referring to some current problem. And Wallach or Rasala or Alsing would reply, "You mean *you* could be in a lot of trouble, right, Tom?" It was Friday, they were going home soon, and relaxing, they could half forget that they would be coming back to work tomorrow.

In the mornings, if West took Alsing into his office, it was usually to ask such questions as: "Is this real, Alsing?" "Is this gonna work, Alsing?" Alsing called these episodes "time out to worry." In the evenings, however, West could ask Alsing those same questions with a wry smile, as if to say, "Weird trip we're on, eh, Alsing?" and the question of whether this project was "real" would usually lead to a little mutual rumination on the meaning of the word itself.

Alsing sometimes encouraged members of the Microteam to visit West's office after six; they would enjoy it, he promised. None seems to have taken the advice, but Alsing was right. At that transitional evening hour, before hurrying away toward his farmhouse, West would leave his door ajar, like an invitation, and leaning back, his hands fallen still, he would entertain most any visitor.

But on the evening after West heard the news about North Carolina, nightfall brought no relief. The game of what's-the-earliest-date-by-which-you-can't-prove-you-won't-be-finished scheduling had been switched on West. Or he had switched it on himself; there was no essential difference. The debugging wasn't going well, and West had a wild air, as if his office were a cage.

He started out talking with his hands in his lap, keeping them busy by twiddling his thumbs. His hands kept getting away, though. He pushed back his hair. He drove his index fingers up under the bridge of his glasses. He made fists and his fingers exploded outward. "I was selling insurance before. Now it's the big crap-shoot, all on one number. Now I gotta get way out on the edge, full of anxiety about the company falling into a revenue hole if it's not there by April — ten thousand jobs on the line, and I gotta pretend everything's fine — and there's a lot of pressure in that. I can't afford to appear too scared around here. I don't talk about it. In the first place, no one outside of here is interested, and I can't talk to people around here and say, 'Here's how I'm manipulating you.'

"Carman says the company's in a lot of trouble if it's not there by April. Suppose I quit? I could just say, 'Fuck it,' and go. . . . I'm not gonna do the next machine. I'm gonna give somebody else a chance to fail. I'm gonna get totally out of computers."

West stopped once in this long, unusually bitter monologue to say: "No, we wouldn't have a disaster. We'd back and fill." He seemed to be saying that he had to *make* himself believe that dire consequences would ensue if his team missed its deadline. In a moment, his reason seemed clear. "It just gets so scary I can't even talk about it. I may have to take Rasala into the lab and work on that thing myself twelve hours a day. When you're doing that, you carry it with you everywhere you go. At some level purely me faces me. I better be up here, just crackerjack every morning, fixing that thing. Every morning I gotta wake up and think, 'Oh, Christ, can I do it?' " West took hold of the arms of his desk chair, as if he were about to get out of it and head for the lab right then. "Rocking back here in my chair and talking about doing it is one thing, but it makes me worry. It gives me a nauseous feeling, because I'm not doing it."

West didn't go into the lab that night or on the working days. One Sunday morning in January, however, when the team was supposed to be resting, a Hardy Boy happened to come by the lab and found West there, sitting in front of one of the prototypes. Then, one Sunday, West wasn't there, and after that, they rarely saw him in the lab, and for a long time he did not even hint that he might again put his own hands inside the machines.

Whether West accomplished anything or not on his Sunday trysts with Eagle, no one ever knew. But some time after he called an end to them, he said: "We're way beyond what any one person can do. It's too complex."

A senior member of the Microteam had wanted to create special microdiagnostic programs for Eagle, and West had turned him down, thinking that the regular, higher-level diagnostics would suffice, as they had for debugging Eclipses. Now he relented. Later, the microdiagnostics would become very important. Opinions varied on how much they assisted then, but it seems that they did help the Hardy Boys surmount the first awful barrier: how to fix the machine enough so that they could really fix it.

For a while West continued to take Alsing aside and ask him if they really were making progress in the lab. Gradually, even those queries ceased. Alsing, who often had the feeling he was watching a movie, believed that a scene ended there. His leading man had finally mastered restraint, and in the running commentary Alsing was keeping through me, he recorded: "This week Tom finally gripped the arms of his chair and decided to trust Rasala."

When West described the other engineers on his staff — Wallach and Alsing — he boasted about their technical attainments. Of Ed Rasala, he said: "His biggest strength is, the guy does not give up. He's in there when most other people go belly up." Rasala was West's trustiest lieutenant.

From the top of his head, the dome of it bald, to his feet, usually shod in construction boots, Rasala looked big. He wore a thin mustache and beard. He delivered firm handshakes and hearty hellos. "How ya *do*-in?" he'd say. "He talks very fast," observed Alsing. "It's almost a speech impediment. Sometimes I feel like giving him a shot of Valium and then telling him, 'Okay, Ed, say it again.'" It was odd, but sometimes, even when just saying hello, Rasala's voice carried a sarcastic sound, as if he was mocking himself for what he was saying.

On an evening at the end of January, down in the basement, Rasala walked up to one of the doors labeled RESTRICTED AREA, unlocked it, and led the way in, down a corridor, around a bend, through another door and into a room with yellowish cinderblock walls on one side and on the other a movable partition. The chamber wasn't much bigger than most suburban living rooms, and it was more cluttered than many. It had a floor of linoleum tile and no ceiling, but instead a tangle of metal braces, heating pipes and, hanging down into the room, many thick black electrical cables. On a gray metal table lay a number of large bound books and looseleaf binders with such labels as EAGLE MICRO SEQ II and ATU PALS. In a gray metal bookcase next to one wall stood published volumes with titles such as *Bipolar TTL Data*

Book. Rasala stood near the center of this chamber of exotic initials, arms folded on his chest, and, nodding across the room, said,
"Here are two state-of-the-art computers, sitting there."

And there they were at last, the object of anxiety: a pair of
Eagles, standing a few yards apart from each other along the
cinder-block wall. They were just two blue metal frames, their
tops about level with Rasala's shoulders. Machines in this state
are said to have their skins off. Inside each frame, exposed to
view — and somehow that didn't seem right — was a shelf full of
what are called wire-wrapped boards — thin plates, each with one
side covered by a profusion of tiny wires. Small cables, flat like
tapeworms, ran among the boards, and farther inside, below the
shelf, hung many bundles of multicolored wire. Oh my, there were
a lot of wires. At the base of the frames, in metal housings, were
the fans that keep the machines from overheating. The fans'
steady roar filled the room.

The two prototypes looked identical, except for their names.
Taped across the top of each frame was a piece of paper. One read
COKE, the other GOLLUM. When the machines were first put together, the team had planned to call them Coke and Pepsi, but
Ken Holberger insisted that one be named Gollum, after the sin-
ister, spidery creature in Tolkien's *Lord of the Rings;* and for some
time that winter and spring, Gollum was in fact the more interesting machine, the one on which they met the most intricate problems.

"They're not too impressive at this point in time," said Rasala.
He meant the machines' looks, not their abilities, though he would
have been right in either case. Equipment much fancier-looking
than the prototypes themselves surrounded Coke and Gollum. On
top of each sat a microNova, Eagle's therapist, with its blue plastic
skins on. A system console — a chair, typewriter and printer —
attended each prototype, and a magnetic tape drive stood next to
Coke. On TV and in the movies, a tape drive often appears in
order to signify the presence of a working computer. That must be

because the reels of tape spin jerkily and thus show that something is happening, but in fact, mag tapes represent some of the least important and slowest parts of a computing system. The real action takes place inside the boards. To get a look at it, you need special tools, and they had them there, of course — boxy little machines called logic analyzers.

The analyzers sat on low wheeled carts, like pieces of surgical gear. They were covered with switches, and each had a small oscilloscope screen; at this moment, on one screen, a strange shape etched in white lines stood frozen and on another danced shifting geometrical patterns. In essence, Rasala explained, these machines were cameras. They took pictures of what happened inside the computer. The analyzers could not look into the chips, but you could hook their probes to the pins of a chip or to wires connecting boards and take several different kinds of snapshots, as it were, of the signals coming and going. The analyzers also contained little memories. When the clock inside Coke or Gollum ticked, every 220 nanoseconds — every 220 billionths of a second, that is — the analyzer would take a picture. It could take and save 256 snapshots during a given foray and play them back on demand.

"It's funny," Rasala said, "I feel very comfortable talking in nanoseconds. I sit at one of these analyzers and nanoseconds are *wide*. I mean, you can see them go by. 'Jesus,' I say, 'that signal takes twelve nanoseconds to get from there to there.' Those are real big things to me when I'm building a computer. Yet when I think about it, how much longer it takes to snap your fingers, I've lost track of what a nanosecond really means." He looked at me. "Time in a computer is an interesting concept."

Then he excused himself. Time in all varieties weighed upon him, and out here it was time to go to work.

We had come in at the changing of the shifts. The second shift, which often became a graveyard shift, was just coming on duty, a little early, as usual — everyone came early and stayed late these days. Ken Holberger, who is short and handsome, was sitting in front of Gollum. Rasala joined him; Rasala looked like a giant beside him. Holberger was Rasala's second in command, and you could see from the easy grin Rasala turned on him that they were friends.

"It doesn't loop on fetch, Ed," said Holberger, in greeting.

Shirt sleeves, jeans and corduroys; mustaches, beards and neat, shoulder-length hair; jogging shoes and hiking boots were in fashion inside the lab. I stood near the center of the room. I heard Rasala cry out: "Whoa. What did you put in for an address? ... I'll give it all F's for the hell of it ... Whoops, that didn't help your float much." Questions would be asked and never answered out loud. An engineer sitting alone at Coke said: "The program blew away on me. I don't know what happened to it." But he seemed to be talking to himself.

Gradually, the scene shifted. Some of the Hardy Boys put on coats and departed. A few others came in. Then Rasala sat down at the central table and took up a sheaf of papers. These were engineering change orders — ECOs. They contained descriptions of recent "fixes" that the team had made in the boards of the prototypes. The ECOs mattered because the engineers were working on different problems in each of the machines. After they fixed a problem in one, they wrote up an ECO so that they'd know how to make the same repair in the other. In his list of seventeen rules for the lab, Rasala had written this commandment: "All boards should be updated each morning so that we can insure that each board is in the same state." Apparently, the crew hated this job. Rasala said they were always putting it off. "For the sake of forging ahead with new problems," he said, "I'm thinking of enforcing the rule."

Rasala rocked back in his chair. "Let's figure out what we gotta do." The others turned toward him. He leafed through the thick sheaf of ECOs, saying: "ATU, ATU, ALU, ALU. And then of course we have the anonymous problem." He made a sour face.

The other engineers had come near — warily, it seemed. Holberger said, "Well, commander, what do we do?"

Rasala looked at them from over the tops of his wire-rimmed glasses, and with a faint, Westian smile, he held aloft the sheaf of ECOs.

Josh Rosen, a black-haired young man said: "Oh, no. Not me. I did that last night."

As Rosen turned away, Rasala stared at him — a lingering gaze, his eyebrows slightly lowered. Then he lifted them and looked at Holberger, who said, "Okay, I'll do the ATU."

Rosen was over at Coke. The others — Rasala and Holberger and a technician — sat down before a long bench on the opposite wall. Each laid a wire-wrapped board down in front of himself.

On one face of a board lay neat rows of the little boxes that housed the chips. The other side, though, was pandemonium. Snarls of thin wires were attached to the hundreds of pins that came out of the chips and extended through holes to these undersides of a board. The wire-wrapped boards rendered a vision of the Jekyll-and-Hyde phenomenon. Each engineer laid his down with the face of Hyde up, and each focused a high-intensity lamp on it. Holberger would pick up the board he was working on and peer into the spaghettilike maze as if he were nearsighted. Rasala left his board on the table and bent low over it. They all followed the same careful routine: locate the pins from which a wire is supposed to be detached; make sure these are the right pins by testing with an ohmmeter; unwrap the wire gently with a small, hollow-tipped tool; throw that wire far away; locate the pins to which a new wire should be attached; wind the new wire on; test again with the ohmmeter; and go on to the next ECO.

About ten minutes went by in silence. Then Holberger said, while peering at his board: "I've got a two-A here. Is that meaningful?"

"Yeah," said Rasala, but he didn't even look up.

Half an hour went by. Rasala pushed back his chair, lifted his glasses, and rubbed his eyes. "The thrill of debugging the machine," he said to me. "We do this ourselves, rather than have a wire person do it. You're so prone to make mistakes with this. We may be a little more cautious. Then again" — he pulled himself back to the bench — "we may not be."

The fans in the two Eagles droned on behind the bent backs of the engineers-turned-technicians. In a while, Rasala turned to me again, holding up before my eyes a pair of tweezers. I looked closely. The tweezers held a tiny strand of wire. "One of the problems," he said. It was awfully easy, he explained, to drop a sliver of wire this size in among the thick nest of wires on the back of a board. And if that little fragment should get lodged in the wrong place, it could cause the machine to commit failures of the sort known as "flakeys." Debuggers could spend hours, even days, finding the cause of such a problem.

I said, "It's like surgery."

"Well, not quite," said Rasala. "Most problems we create are repairable."

I laughed.

But he wasn't finished. "Most surgery isn't," he added, and went back to his labor.

When it came to working on the machine, Rasala had no truck with incomplete thoughts.

"I come from a conservative background, from the beer-drinking generation — I'm into sports, my wife's into being a housewife — so I don't have dreams of grandeur fundamentally about going out and building the next evolution of the computer."

The first few times I talked to him, Rasala described himself in long rapid sentences like that one, and he spoke them with an edge in his voice that seemed to say, "You probably think I can be sized up like this."

"I'm a poor Polish boy, I come from the slums of New York, I have a high-school equivalency diploma and would rather play softball," he said. In fact, as he later told me, his father came from

Poland, and Rasala grew up in an apartment in Brooklyn. But it wasn't the slums, and what he really remembered was a home where you never went hungry and you knew from an early age that you were going to college. Both of his parents worked. His father, a waiter for Child's Restaurants who read a great deal on the side, taught himself from a book to fix TVs and made a part-time business of it. "Both my parents are very smart," said Rasala. His older brother was "a math whiz" and wound up valedictorian of his class at Columbia University. "Pretty good," said Rasala, "considering my mom has a high-school equivalency diploma and my dad never went past fifth grade."

Although not a bad student, Rasala had the misfortune of following his brother through school, and from the elementary grades on, his teachers never stopped reminiscing about his brother. "From grade school on, it was always my rap — I never lived up to my potential," he said. He didn't sound bitter; he seemed to believe it was a matter of fact.

Rasala went to Rensselaer Polytechnic Institute and studied electrical engineering, including logic design. But this was during the early sixties, and he never laid hands on a computer until he went to work at Raytheon, right after college. Rasala never had the feeling, which others in the Eclipse Group remembered, of having his true calling suddenly revealed to him the moment he touched a computer. And it took him seven years, he said, to recognize "the opportunities and pleasures of working.

"To me, it was more important to play softball, take karate lessons, and play hearts at lunch." At Raytheon he carved out for himself an undistinguished, comfortable niche, doing routine assignments. When he finally did get an assignment that he could not fulfill by looking up answers in books, he worked hard and succeeded, somewhat to his own surprise. That project whet his appetite. But he knew he had established a reputation for being "an average guy," and he did not think other interesting assignments would come his way at Raytheon. To shuck that reputation, he decided he had to start over at another company.

In a trade paper, he saw an advertisement for Data General — not one of the early, brazen ads, just an ordinary listing: "Engineers Wanted." On a whim, he drove out to the company's headquarters and said, in essence, "Here I am." Engineers and executives, including Carl Carman, interviewed him on the spot, and Rasala was impressed. "At Raytheon, I never even *saw* someone at Carman's level." Data General offered him a job that very day and he accepted their terms without so much as an argument. "I wasn't very good at bargaining, I guess." And he went to work, not on the job he thought he had been hired to do, but on a piece of the hardware for the first Eclipse. "I was had," he recalled, and he smiled.

When Rasala arrived, all that remained to be done on his part of that Eclipse hardware were finishing touches. West, who was just another engineer then, noticed Rasala, however. He liked the dogged, straightforward way in which Rasala went about finishing up. As Rasala himself observed much later on: "With Tom, it's the last two percent that counts. What I now call 'the ability to ship product' — to get it out the door." Rasala looked at me squarely. "And I may not be the smartest designer in the world, a CPU giant, but I'm dumb enough to stick with it to the end."

West called himself "a mechanic" — one who could take the ideas of engineers technically brighter than he and make those ideas work; he thought he had spotted that quality in Rasala, too, and he had promoted it. West groomed Rasala as a leader of hardware teams. Eventually, he put Rasala in charge of the hardware for the biggest of the 16-bit Eclipses, the M/600. As often happened, the pace of that project became frenetic. It took a year of increasingly intense labor from Rasala, and when it ended he felt tired. He was looking around for something easy to do when West asked him to manage the development of Eagle's hardware. Rasala declined.

In his office, the door closed, West schemed. "How am I gonna get Rasala to sign up?" he asked Alsing.

Unlike Wallach, Rasala wasn't a student of advanced architectures. Rasala wasn't worried about Eagle's supposed inelegance; he was just tired. Offering him a chance for vengeance wouldn't work, because Rasala had been pounding out the M/600 during the EGO wars and he didn't have anything to feel vengeful about. Rasala was a hard one, West said. Bringing him to the signed-up condition was mainly a matter of persistence. West presented Eagle to him as a test of strength. Could they get this machine out the door on time? Again and again, he told Rasala that the company needed this machine desperately. Rasala never actually said, "I'll do it." One day, it seemed to him, he was just doing it.

For months afterward, he would come home at night and his wife would ask him, "How was your day?"

"It was terrible," Rasala would tell her. But as he went on describing the day's events, his wife noticed, he became increasingly excited.

"Maybe it's masochism," Rasala said. "But I guess the reason I do it fundamentally is that there's a certain satisfaction in building a machine like this, which is important to the company, which is on its way to becoming a billion-dollar company. There aren't that many opportunities in this world to be where the action is, making an impact." It struck him as paradoxical, all this energy and passion, both his own and that of the engineers around him, being expended for a decidedly commercial purpose. But that purpose wasn't his own. He had enjoyed his years at Raytheon; life had been pleasant there and he had been an easygoing fellow. Now, in the Eclipse Group, for several years in a row he had been working overtime without extra pay in an atmosphere that was decidedly not easygoing. Why had he made the switch to Data General and now signed up to work on Eagle?

Rasala said, "I was looking for" — he ticked the items off on his fingers — "opportunity, responsibility, visibility."

What did those words mean to him, though?

Rasala shrugged his shoulders. "I wanted to see what I was worth," he explained.

More than anyone else in the Eclipse Group, Rasala borrowed expressions from West: *fundamentally; basically; the win; some notion of; quick-and-dirty;* the ubiquitous *canard;* the long, drawn-out *and,* as in, "So and so says such and such, ahhhhhnd, it's a canard." But Rasala tended to qualify and expand, as if he feared ambiguities, whereas West was inclined, in his daily commerce, to the short, forceful sentence. West's words and phrases just didn't sound the same on Rasala's tongue.

The team was having problems with one of the support groups. They believed that this group was slowing Eagle's debugging; this group wasn't trying hard enough — which was to say, they were working just normal hours on Eagle's behalf. They believe Eagle will fail, Alsing thought; he could see it in their eyes, he said. West stood in a corridor discussing the problem with Rasala. "I'm gonna detonate those guys," said West, in a flat, calm voice, as if he were planning a business trip. Then he promised to "waste 'em." He added, "We'll string them up by their toes."

Months later, looking back, Rasala decided that when West took on people in other groups and "beat them up," as this sort of activity was called, he did so in a purposeful way, one that was usually "not so much mean as intimidating." In one fairly typical case, a support group missed a promised deadline. At a meeting, West questioned the group's leader. The leader said that he hadn't received a certain piece of equipment yet. West would not let the matter rest there. He wanted to know why it hadn't been shipped and what was being done about it. West was "always pushing," Rasala noticed. But the act, Rasala would come to believe, was never as dramatic as the language West used to describe either what he was going to do or what he had actually done. For a long time, however, Rasala took West's descriptions as a fairly literal model of how to contend, and when Rasala confronted a troublesome support group, he beat his fist on the table and shouted and made open threats. I saw him after one such encounter, returning to his cubicle. He was red. "I don't want to beat people up," he said, throwing his hands down. "I don't want to be a bad guy. I just want to get something done." He took some deep breaths, then proceeded to explain that it really wasn't the support group's fault. Eagle wasn't their machine, they couldn't be expected to work as diligently on it as the Eclipse Group.

Bob Beauchamp remembered, by way of drawing the contrast between West and Rasala, receiving a set of orders from West. "Do this," West said to him, and in West's tone of voice, Beauchamp heard, "No ifs, ands or buts, no arguments, no commenting, no nothin' — just figure out how to do it and get it done."

"Did it work?"

"Yeah, it worked," Beauchamp said. "When Tom said do something, people did jump." He went on, "Rasala was practicin' that, I think. He could bark too."

Beauchamp grinned. He and Rasala were good friends now. Before they were, early in the project, Beauchamp was standing around with one of the Hardy Boys one evening, talking casually, when Rasala came out of his cubicle and, looking them up and down, said, "Don't you guys have work to do?"

"What the hell?" Beauchamp thought to himself. "It's after working hours." So he said to Rasala, "No."

To Beauchamp's surprise, Rasala didn't say any more. He just walked back to his cubicle.

"Then I knew," Beauchamp said.

Rasala participated in West's campaign to create the illusion of a machine before one existed, and on the whole Rasala endorsed the strategy. But he worried. Clearly, he could not make West's approach his own. "We're getting the rest of the company geared up now, and we could fall flat on our face. This whole thing scares me, because we're pushing." At West's insistence he had made up a debugging plan that would bring Eagle in by April. Then he made up one that would bring the machine home by May. In mid-March, when Rasala knew that they wouldn't come close to that second mark either, he went to a meeting with West and heard his boss insist to a number of people in others groups that they would have all the bugs out by May. "Now May one is the target and we're not gonna do it," said Rasala when he came back. "It's sort of a poker game. Everyone's bluffing, but everybody seems to know the rules of the game, and I don't know those rules yet." He continued, "I strongly believe in supporting my boss, because I know he'll support me. So I try to be consistent with what Tom is saying, one, and two, I try to be as honest as possible."

One day Rasala bought a self-help manual that promised on its cover to hold the secret to understanding one's boss.

Just a few minutes after I first met him, Rasala told me, "I'm not a bright guy." I wrote that down in my notebook, and for the rest of the evening, he kept saying: "But I'm a dumb guy. I must be, you wrote it down." Rasala was far from dumb, though. He was just somewhat puzzled and, at the same time, bent on self-improvement.

When Rasala said something earnestly — if they failed, it would be his fault, no one else's, he told me, for instance — you never felt that he meant something else. Somehow he found the time, after working what usually amounted to a ten- or a twelve-hour day, to drive into Boston one night a week and take a course in a new programming language. He was going to fill in the gaps in his technical education, and he would not skip a class, not for a snowstorm or even a ball game.

Most engineers, I think, consider themselves to be professionals, like doctors or lawyers, and though some of it clearly serves only the interests of corporations, engineers do have a professional code. Among its tenets is the general idea that the engineer's right environment is a highly structured one, in which only right and wrong answers exist. It's a binary world; the computer might be its paradigm. And many engineers seem to aspire to be binary people within it. No wonder. The prospect is alluring. It doesn't matter if you're ugly or graceless or even half crazy; if you produce right results in this world, your colleagues must accept you. It's an exciting environment to contemplate; you can change the way people think if you can provide the right reason, and you can predict the way in which others may change you. Since there are only right or wrong answers to questions, technical disputes among engineers must always have resolutions. It follows that no enmity should proceed from a dispute among engineers.

Clearly, West believed in these precepts — designs were right or wrong; engineers were winners or losers and those judgments had nothing to do with whether he liked them personally or not. But there were many reasons to doubt the reality of this binary world. What had the EGO wars proven but that talented engineers can dispute narrow technical issues and never come to agreement? Amazed at the alacrity with which West sometimes consigned engineers around him to the status of winners and losers, and convinced that West was usually right, Alsing nevertheless always wondered to what extent West made his own judgments come true. Rasala had always believed that a technical argument between two engineers couldn't lead to hard feelings. But he had recently lost his oldest professional friend because of what he had imagined to be a technical disagreement.

An engineer was supposed to be eager to advance in his company's hierarchy, and Rasala was, but he had other dreams. Some years before, he had taken a trip to Jackson Hole, Wyoming, and ever since, he had wanted to go back there and do something simple: he thought he'd like to open a grocery store in Jackson Hole someday. To Rasala, many of the young Hardy Boys seemed hip and cultivated in ways that were foreign to all his previous experience of engineers. But he did not for that (or any other) reason disapprove of them. And he didn't seem to envy them or want to change his ways and be just like them. He talked about them with plain curiosity, as if he were a traveler in their country.

One evening, over beers, Rasala complained about some insipid movie recently shown on TV. He compared this stinker to *The Prime of Miss Jean Brodie,* an example of a film he'd liked. Actually, he had admired the protagonist most of all. "I don't know why," he said. "She's romantic, foolish, unrealistic — everything an engineer's not supposed to be. But I like her." He gave a short nod, as if to say, "This decision is final." Then he said it again. "I like her."

Rasala wasn't West and he wasn't dumb either. He seemed to be in transit.

When the hardware team started to design Eagle, Rasala opened a sort of diary. It reads like the journal of a frontiersman's travails, except that instead of wolves, hostile Indians and busted wagon wheels, the reader encounters the indecisiveness of logic designers, an affliction that causes a designer to "spin" from one possible approach to another; the regular breakdown of the computer they were using to work on their computer; and the tendency for all bright young engineers to redesign and redesign in search of perfect unattainable solutions. In the diary, one delay follows another. Schedules always slip. The last entry, recorded several months before they stopped designing, reads, "Overall, things look lousy."

This was the biggest job Rasala ever had, and he was just finding his way around in it when he wrote that journal. I got the impression that a few engineers doubted his suitability, on the grounds that others were more brilliant than he. Rasala himself said: "I'm an implementer. I'm not gonna go out and invent anything. But making it work is fun. It's something I think I do reasonably well. I don't have Wallach's knowledge, I'm not on top of the architecture per se, but I'm a good designer, I think, and I'm a better debugger."

The sheer physical complexity of the machines demanded that the debuggers plod, and plodding did not always come easily to the quick, bright Hardy Boys. Were signals just barely making it to their destinations between ticks of the computer's clock? If so, they had better go back and work on problems they thought they had solved. Was there a lot of noise inside the machines? Noise, as Rasala explained it, is what makes your TV go haywire when you turn on your blender — stray voltages that are propagated through the machine; and too much noise fouls performance. Noise is not the most interesting of problems, but coping with it requires experience and imagination and a certain doggedness. On one occasion, the Hardy Boys found that while one board was failing, another one, identically wired, worked right. The problem, they discovered, lay in a single chip. They wanted to throw that chip away, put in another, and go on. But Rasala would not allow it. The suspect chip might not be defective; it might just be working a little more slowly than most of its kind. In that case, they would have a real problem to fix. It might never show up again in the lab, but when the time came to mass-produce this machine, it would almost certainly come back to haunt them. Some slow chips were inevitable, and they had to provide for them.

In the main, Rasala's technical role was to play the brake. When the Hardy Boys complained loud and long about updating boards every day, he bent the rule, but insisted that they perform this crucial, unpleasant task every Saturday. When he put them on two shifts, he made a point of working the better part of both. He was a good West Pointer; he led by example.

Rasala took a week off from the debugging that summer and spent the time building a porch on his house. I came by one day to give him a hand and got a dose of his style of management.

He lived in a fairly new house, of quasi-Colonial design, located in one of the many residential neighborhoods that had sprung up in the area more or less in rhythm with the growth of the numbers on the bottom line of Data General's annual reports. The numbers had grown faster than trees do. His house lay so close to headquarters that he could have ridden a bike to work, although, given the traffic nearby, it would have been a dangerous commute.

While working with Rasala, I made a careless reading of a carpenter's level and we consequently made the frame of the porch just a shade cockeyed. Rasala stood back and stared at the frame for a while. Then he said, "Ship it."

He teased me about the mistake, in a friendly way. He also prodded. When I arrived, he said he expected to finish the entire job that day. Around three in the afternoon, I said I was tired. "Tired?" said Rasala, reaching for a tenor's high note. "Tired? You're not tired. How can you be tired? You mean to say you're tired already?"

Somehow, I felt I had to keep working. An hour and a half later we still weren't finished, and then I really felt tired and said so. Rasala let go on the instant, brought out some beer, and declared the day a success.

For most of their time together, most of the Hardy Boys got along well with Rasala. He teased them and he prodded them, and they did the same, both to him and to each other. One evening just before going home, Holberger made a minor, careless mistake. Discovering it, Rasala told the rest of the crew, "I hope you're gonna leave Holberger a nasty note. He'd leave you one." Rasala liked a contentious atmosphere, a vigorous, virile give-and-take among himself and his crew. "Smart, opinionated and *nonsensitive,* that's a Hardy Boy," he declared.

Above all, Rasala wanted around him engineers who took an interest in the entire computer, not just in the parts that they had designed. He said that was what was needed to get Eagle out the door on time. He wanted the Hardy Boys to bind into a real team, and he spoke with evident frustration of engineers who were reluctant to work on boards that someone else had designed, who felt comfortable only when working on their own. Josh Rosen, he believed, was one, only one of those who felt this way. "Rosen designed the ALU and he wants to work on the ALU and nothing else," Rasala complained. "But I have a good confidence level on that board. It's not the first priority now. I need him elsewhere." Much later, Rasala would say, "I realize now that Josh wasn't as comfortable with the ALU as I was." Back then, however, his feelings were different.

It seemed that he had given up trying to win Rosen over. When he spoke to Rosen, Rasala's voice had an edge to it, like the one it sometimes had when he was describing himself. He teased Rosen from time to time. "Your ALU is not fast enough to keep up with this blazing machine, Josh," he told him one night in the lab. Rasala often spoke that way to Hardy Boys. The difference was that the others usually answered him back in kind. Rosen, on this occasion, just turned away.

The days of the debugging wore on. In March West said, referring strictly to the debugging, "Most of the fear is gone now." He was speaking only for himself, however. The team had passed through the first sharp fear. But they had designed the machine much too fast for prudence. It had features that none of the group had dealt with before. At this stage none of them dared claim to understand in detail how all the parts worked and fit together. Sufficient cause for worry about the debugging remained.

Alsing thought that even under the best circumstances, several kinds of fears inevitably attend a debugging. One was the fear of "the big mistake," the one that would be discovered late in the game and would require a major redesign — and with it, perhaps, a fatal delay. There was "the flakey fear": that they had designed Eagle and were debugging it in such a way that it would never be reliable or easy to build in large numbers. Rasala had that one fairly well in hand. But there was also "the bogeyman fear." "Just something dark and nameless," said Alsing, "that the machine just won't ever work." West said: "It's the infinite page fault you didn't anticipate. The bogeyman is the space your mind can't comprehend."

Once in a while, Alsing said, he suffered a form of this last anxiety. "Maybe these guys writing the microcode are really bullshit artists, so full of shit they don't even know they're full of shit. Maybe this whole thing is bankrupt." Such thoughts came to him rarely, usually at night. They vanished when the sun rose. Probably they stemmed from the fact that Alsing wasn't writing this code himself or reviewing it closely. Rasala once remarked, "Yeah, the further you get from doing it yourself, the more demons you see."

Gradually, Rasala came to own (having acquired the title to it from West) main anxiety over the machine. Anxiety became just something he lived with, like a bad back. For a while, I engaged in a ritualized conversation with him.

"How's the machine, Ed?"

"Ahhhh, la machine," he would say. "Let's see. How is it?" He would pull his chair up to the chart of whatever debugging schedule was currently hanging on his cubicle wall, and usually wound up explaining why they had fallen behind this schedule, too.

One day he answered the question differently. *"The* machine. It's what everyone calls it. That's the whole thing — to build *'the* machine.' " When Rasala heard that phrase, he thought of a movie called *Duel,* which he had seen on television a couple of years before. In it, the hero is chased by a trailer truck, for no apparent reason. Throughout the movie, as Rasala remembered it, neither the hero nor the viewer ever gets to see the truck's driver, if there is one, but only the front end of the semi looming like a huge evil face in the hero's rearview mirror, always threatening to drive the hero off the road, always returning just when it seems that the hero has escaped from the thing at last. "My favorite movie!" Rasala said.

His five-year-old son, the older of his two children, soon found out his father's weak spot, and now when he got angry at Rasala, the child would say, "I hope your machine breaks, Daddy."

At night, when the last Hardy Boys got ready to go home, they would leave Coke and Gollum running one of the many diagnostic programs, a long list of tests that the machine had to pass before it could be called truly functional. Usually they'd leave the prototypes running a test that they had already mastered. At home, in the middle of the night, Rasala would sometimes come awake with a start. He would not be conscious of having dreamed.

He would awaken and find himself wondering if one of the machines had stopped working for some new, unknown reason. Or he would wake up thinking about the latest failure, the one whose cause they'd been looking for a whole week and still hadn't found. The bogeyman — la machine — was there in his bedroom.

8

THE WONDERFUL
·MICROMACHINES

TO ALMOST EVERYTHING they touched, the Microteam attached their prefix. The office that four of them shared, sitting virtually knee to knee, had a sign on the door that said THE MICROPIT; the room in which they held their weekly meeting was the micro-conference room. They gave out microawards and Carl Alsing had his microporch. One of them owned a van, which became the microbus. That winter several of them would go out riding in it on Friday afternoons, when West held his own weekly meetings with his managers. Then, in the first warm days of spring, they created the outdoor microlounge, to which they now repaired on those Friday afternoons.

Near one corner of the back of Building 14A/B, down a steep, man-made, earthern embankment, lay a narrow slice of woods. It must have been a pasture once. An old stone wall ran through it. Several flat rocks had been removed and propped against the wall in such a way as to fashion the crude seats and backs of micro-chairs. The sky promised rain, and the leaves not yet being out, the lounge lay in full view of the camera mounted on the corner of the roof high above, but the several Microkids who gathered there didn't seem to notice the weather or the camera. Their mood was altered. Reclining, they talked about computers and other social issues.

"Our nightmare is that you keep getting faults."

"We have to do this project on the fly."

"Yeah, we're all flying."

"No, if you page-fault you have to look for it, but if you can't find it you have to page-fault to find it and you can't page-fault until you find it. That can happen with a stack fault, too. It's a basic kind of crock."

"Sure. There's always a way to kill a computer."

"We should build a trap that'll let us in, for fun, at the end."

"That's a nice concept, leaving a trapdoor for later, but a real purist would want to get into a machine built by someone else."

"Well, there aren't too many machines in this place that we can't control."

"What does it mean to understand a computer?" I inquired.

"Knowing where all the electrons go?"

"Maybe the electrons aren't in the computer at all. They just turn on the lights."

"Electrons are mathematical abstractions. So who can talk about electrons? We speak about electrons loosely."

Everybody laughed and laughed, until, it seemed, everyone forgot what they were laughing about. One Microkid started talking about a model of computer that was ten years old. He said, "They're really *ancient.*" Uttered there, beside the crumbling stone wall, the statement sounded odd.

"I've been here since seven," said one of the team, apropos of nothing.

"I've been here since January," said another.

A couple were discussing solar energy, the military-industrial complex and education. Engineering school was awfully specialized; meanwhile, the liberal arts were in decline. "Liberal arts, they're not economically viable," said one.

"Does that mean you should restructure education or restructure society?" said the other.

"You should restructure society. That's for sure."

It was a pleasant afternoon. Looking up toward the TV camera, I noticed a large bird's nest in a nearby tree. Trying to catch their mood, I said: "Look at that bird's nest. Who do you suppose lives there?"

Right away, in a tone of voice expressing something like appreciation, one of the Microkids looked up at the nest, nodded judiciously, and declaimed, *"Carl Alsing."*

Part of the Microteam displayed a tendency to keep odd hours and play zany games. You'd expect that. Compared to that of the Hardy Boys, who were now bound by necessary schedules to the actual machine, the Microteam's job had an air of the ephemeral. In the main, they could write their code when they pleased, as long as they did it on time, which is to say, very quickly. The code, the go-between for language and machinery, is in itself odd stuff — a witches' brew, many novices feel the first time they see it work. "It takes people with devious minds," Microkid Dave Keating contended. "Almost all of us fall into that category. We're not quite system bombers, but we're shaded that way somewhat."

"I hire Wests. Alsing hires Alsings," said West. But both West and Alsing had chosen recruits with fine academic records, unlike their own. And in other respects Alsing had shown eclectic taste. He hired a woman — and women were scarce in hardware design — as well as a former rock-and-roll musician. He picked some shy ones and some much given to practical jokes. At least three of his choices would turn out to be, when given the chance, fine engineers both of hardware and of microcode. And when he had hired Chuck Holland a couple of years earlier — this and the fact that he trained Holland was one of Alsing's crucial contributions to the project, as it turned out — he did so largely because of a piece of sculpture. "I'm working on this thing," Holland had said, when in his interview, Alsing had asked him what he did for fun.

"Tell me about it," Alsing had said. So Holland had, and by the time he had finished, Alsing had felt sure that Holland could write microcode. "It was my clue that he could make something intricate work."

From a distance Holland's sculpture looked like a rectangular cage, and closer up, a gallery of spiders' webs. It stood roughly five feet high. Thin steel rods attached to each other at odd angles made up the four outer walls of the cage. The top made a sort of funnel, somewhat like a coin box on a bus — four sloping planes that narrowed down around a little hole. Rather shyly, Holland dropped a handful of silvery steel balls into this funnel, and the sculpture went to work.

Within it, in puzzling shapes, lay a complex of tracks made of two thin steel rods. A ball dropped through the hole at the top. An entire section of the cage seemed to move. Inside the webbed structure a section of the labyrinthine track swung down; the ball dropped onto it; the seesawing piece of track rose back to its starting place, blocking the next ball for a moment. Then it swung down again. The second ball dropped. In a moment, the thing was full of steel balls and all of it was softly clanging, like a city awakening at dawn. It made you blink to watch it work. The balls, descending, seemed to cross each other's paths. Now one seemed to be in the lead, now another. They made sudden sharp turns, zipped from one side of the cage to the other, disappeared from view behind a maze of fretwork, traveling ever so slowly, and reappeared as silver blurs.

"Do it again."

In fact, there was really just one track and the balls went down it one by one. The track was also discontinuous. A ball rolled down one stretch of it, balancing between two rods, and the rods of the track widened just enough, at just the right spot, so that the ball would drop through from one section of track down onto another section. "There are a lot of fine tolerances here. The track got a little rusty once and then it wouldn't work," said Holland.

"At first I just wanted to make it work. It was an engineering feat for me. Then I decided I wanted to hide the track, but then I decided the track's kind of neat so I let some of it show. I wanted it to have some meaning. I have my own ideas of form. Planes and so on." He ran his hand across one section of the outer walls, as if stroking an animal's fur. "These wires cut this plane, and I just think that's really neat." It had taken Holland months to build this thing, taking it apart, putting it back together, taking it apart again.

His father was an engineer; Chuck Holland had wielded screwdrivers and pliers for as long as he could remember. He had met his first computer in high school — and it was an old IBM, of course. Good-looking, trim, neat in his appearance, he had a wild-looking smile, which he did not often feel like showing during the season of Eagle. He did not approve of the way the group was assaulting diagnostics, and he pleaded, without success, for a more diplomatic approach. He seemed quiet, pleasant, shy — all in all somewhat like Alsing, easily overlooked — and he felt that he had been, thus far in his career.

Holland organized the microcode for Eagle. Each microinstruction of a microprogram is a string of 75 units of high and low voltages, represented by 0's and 1's when written down. Holland divided this standard 75-bit string into standard chunks of several bits each, each chunk being called a field. Each field contains some number of unique combinations of 0's and 1's and affects discrete parts of the hardware. Each unique combination is what Alsing called a microverb. Holland and Ken Holberger invented definitions for every possible microverb. The result was a sort of dictionary. Adding versatility to its contents was the fact that certain microverbs could, for instance, say one thing to the ALU and something altogether different to the Microsequencer.

The possibilities for creating microcode full of internal contradictions were virtually unlimited. Holland had to guard against that. He drew up rules — of grammar, as it were — to make sure that the code and hardware would fit together and to prevent each microprogram from interfering in some subtle way with any other. He saw to it that every microcoder consulted the others when making a change in a microverb. The microverbs and rules for their application he put into a book, called UINST. *U* stands for "micro" and *INST* for "instruction set." The Microteam called it their bible. The Hardy Boys called it "a Sears, Roebuck catalogue" and "the Microteam's wishbook."

UINST became a battlefield. Its contents changed every week. Holland and his troops would make their changes and the Hardy Boys would look at UINST and say, "There's no way we can do this function and this function in hardware." The two sides would argue and work out those problems. Then the Microteam would discover something else that was hard to do in microcode, and deciding that it should be done in hardware, they would insert in the next issue of UINST this wish for a change. The Hardy Boys were on guard. They'd scan UINST carefully, looking for new Microteam mischief, and finding this new item, declare, "No way we can do that in hardware." And it was back to bargaining again.

Control seemed to be nowhere and everywhere at once. It was an almost tangible commodity, passed from hand to hand down the hierarchy of the group, and everyone got some. West gave Alsing responsibility for getting the microcode done on time. Retaining ultimate authority, Alsing let Holland assume almost complete technical command; and after establishing general rules and while keeping an eye on them, Holland gave each Microkid virtual reign over a portion of the code.

There were hundreds of basic instructions to be encoded in thousands of microinstructions. All of the Microteam worked on recoding the old 16-bit Eclipse instructions. There were the microdiagnostics, or "the Wringers," as they called them. They worked at all hours, and sometimes when the Hardy Boys needed a new section of code in order to move on with the debugging, they worked around the clock. Every member of the Microteam knew weeks of intense pressure, which would gradually subside, and then rise again.

Jon Blau said: "You have narrowed your field of vision to a small little world, trying to make it a vision of your own mind, and that's the kick — getting control of something. Maybe that for me describes the kick — that I can be totally in control of XSH-zero. That's the name of a signal, which is the electrical embodiment of an idea. It was wrong. This morning I found an easy way to fix it." It did not always seem healthy to Blau. "When you're concentrating on that little world, you leave everything else out." But he had to admit, he was enjoying himself. "That's the big kick," he said. "That the guys with the purse strings are trusting a bunch of kids to come up with the answer to VAX. That's what bowls me over, that they haven't just put us in a corner somewhere, doing nothing."

The whole process looked strange, confusing, improbable, even to some in the Eclipse Group. But although the group as a whole was missing deadlines, the process seemed to be working. They were developing the hardware and microcode more or less in concert, and the Microteam was holding up its end. Although most of them were only beginners, when they delivered their code to the lab, they usually did so on time; and they made mistakes, of course, but far fewer than Alsing had expected.

One main reason why they did not fall into an enthusiastic chaos was Holland himself. He helped Alsing choose the team. He organized their work. He reviewed all of it carefully. He helped mediate the battles over UINST. He wrote some large chunks of code himself. No one had more skill than Holland at the art of intricate construction, and no one worked harder than he. Nobody had ordered him to do all this. Alsing had made the opportunity available, and Holland had signed up. On the whole, he was happy in this work. "I never had more people around me who felt like they were friends," he said. He, too, felt excitement in the air. But he had some experience. He knew that credit did not always fall to those who deserved it. Would Alsing and others take credit for things that he, Chuck Holland, had done? The fear made him gloomy sometimes. He seemed to be girding himself for disappointment.

Holland did not make all arrangements. Another source of the Microteam's strength was a tool they called the simulator. They relied upon this thing. They loved it. They did not like to think where they would have been without it. And it was Alsing who arranged for the simulator, in an odd sort of way.

In theory, a computer can mimic the behavior of anything. It can do so accurately only if the thing being imitated is thoroughly defined. So computers achieve only partial success, at best, when instructed to simulate the behavior of a city or to foresee the future of a national economy. Computers do well, however, when imitating other machines, including other computers — unbuilt ones that exist only on the paper of an architectural specification. You make the old computer imitate the new by writing a program. This program — the simulator — makes the existing computer respond to instructions just as the contemplated, unbuilt computer should. Essentially, your program translates instructions designed for the unbuilt computer into instructions that the existing one obeys. You can create a simulator that will make an old computer ape the behavior of the fanciest new computer you can imagine. People build new computers in hardware and not in such a program because simulators are slow — the Microteam's for instance, ran more slowly than Eagle was supposed to run by a factor of about 100,000. A simulator makes a slow computer but a fast tool.

Alsing wanted one. He had often wished he had a simulator, for testing and correcting microcode. On every other project, he had been forced to debug the microcode by running it on the prototype hardware of the new machine. But the hardware of the new machine was itself being debugged and indeed could not be fully debugged without the microcode. This made for awkward situations and mysteries in the lab. If the machine failed, it was hard to know what to blame first — the hardware, the microcode or the diagnostic program. For several years now, before every project, Alsing had held essentially the same conversation with West.

"I want to build a simulator, Tom."

"It'll take too long, Alsing. The machine'll be debugged before you get your simulator debugged."

This time, Alsing insisted. They could not build Eagle in anything like a year if they had to debug all the microcode on prototypes. If they went that way, moreover, they'd need to have at least one and probably two extra prototypes right from the start, and that would mean a doubling of the boring, grueling work of updating boards. Alsing wanted a program that would behave like a perfected Eagle, so that they could debug their microcode separately from the hardware.

West said: "Go ahead. But I betchya it'll all be over by the time you get it done."

Simulators were at least ten years old. No great mysteries surrounded them. After some calculation, however, Alsing realized that a program to simulate Eagle would be huge. It might take a seasoned programmer a year and a half to write such a thing, he figured. But Alsing kept these calculations to himself.

Although his intentions were in the main unquestionably gentle, Alsing seemed to have a constitutional aversion to the direct approach. I was visiting him one evening that spring. We were sitting in his living room, when the oldest of his three sons, a handsome, soft-spoken and polite teenager, came to him complaining that all of the TV sets were broken again.

"What did you want to watch?" Alsing asked him.

" 'Charlie's Angels,' Dad."

"Gee, that's too bad about the TVs," said Alsing.

When his son had left the room, Alsing grimaced. "I'm terrible," he said.

He had grown worried that his boys were watching too much of the wrong sort of television — "evening violence and Saturday-morning cartoons," he said. So one night he had gone around the house and had disabled all of the sets. On the theory that his sons would learn more from trying to repair a TV than from watching one, Alsing encouraged them to try to fix the things. He spent many pleasant Sunday mornings working on the sabotaged sets with his sons. His boys were learning more and more about TVs, but they hadn't yet gotten any of them to work. On occasion, however, when Alsing approved of a show that they wanted to watch, at least one set would, as if by magic, suddenly start working.

Who would Alsing get to write the simulator? Who could do it quickly enough for the thing to be useful? On the Microteam there was a veteran programmer, a jovial fellow named Dave Peck. Peck had a raucous laugh; it was his laugh preeminently that drove West to distraction. Data General employee number 257, Peck had performed his full share of round-the-clock programming. He had come over from Software to the Eclipse Group, because Alsing had promised him the chance to manage others for a change. Peck was a fast programmer. "The fastest I ever saw, the fastest in the East," said Ken Holberger.

Peck said he'd been told he was fast so often that he guessed he must be. "But programming's just obvious to me." The problem was that these days, it took him a long time to talk himself into writing a big program — longer, as a rule, than it took him to write the program. Peck could turn out a simulator quickly, if anybody could. (He would ultimately perform some crucial smaller programming projects for the group, about thirty-four in all.) Like Alsing, he had been thinking on his own about a simulator for debugging microcode — just figuring out how he'd go about writing a simulator, in case he should get up the desire to do it. So far, however, the desire had not come.

The group also had another programmer. Among the first of the Microkids to arrive at Westborough was a twenty-two-year-old phenomenon, a fellow with degrees in computer science and electrical engineering and a nearly perfect academic record in those areas. His name was Neal Firth. He liked to program computers. He said: "I may be a little vain. I took a course in college in machine code. It was supposed to be a flunk-out course. To me, it was extremely simple. It's always just seemed extremely logical to me, programming."

Alsing considered the situation. He had two confident programmers — one relatively inexperienced, the other reluctant. Somehow, out of the two of them, he should be able to get a simulator. He put Peck in charge of the effort, but only nominally.

Shortly after Firth arrived, Alsing sat down with him and discussed ideas for a simulator. "There are a number of things you have to do to write one," said Alsing, drawing a grossly oversimplified picture of the task.

"Yeah, I could do that," Firth replied.

They talked about simulators with growing enthusiasm, until Alsing felt they had "a nice little fire going." Then casually, Alsing popped the question: "How long do you think it would take to write this simulator?"

"Six weeks or two months?" said Firth.

"Oh, good," said Alsing.

When he was ten years old, Alsing remembered, he was given a book called *All You Need to Know about Radio and TV*. When he had read it from cover to cover, he really believed that he did know everything about radios and TVs. He didn't, of course, but thinking that he did had given him the confidence to take apart radios and TVs and in the process to learn what made them work. He had that experience in mind when he talked to Firth. "When Neal said he could do it in a couple of months, he was probably thinking back to how long term-projects in college took." Firth, Alsing reasoned, had not been performing make-work projects as many neophyte engineers do their first years out of school. He didn't know what he could not do. "I think that after our little talk, Neal had a picture in his mind that he knew all about simulators now. It was no problem. He could do it over the weekend."

Around this time, Alsing also sat down with Peck. Firth was Peck's responsibility, Alsing said. Peck should somehow lead Firth to complete a simulator in just a few months. So Peck explained to Firth his own ideas about simulators. He laid out the basic scheme for Firth, in greater detail than Alsing had.

Observing the two from a little distance, Alsing got the strong impression that while Peck and Firth liked each other, neither was about to concede that the other was a better, faster programmer. A little friendly competition might work nicely here. Alsing went to Peck and said: "We really need some sort of simulator in six weeks. So, Dave, why don't you do a quick-and-dirty one, while Neal works on his bigger one?" Peck agreed. The contest began.

Every so often over the next month or so, Firth would visit Alsing and tell him that *his* simulator was coming right along and was going to do many things that Peck's quick-and-dirty one would not.

"Oh, good," said Alsing.

Peck got his simulator written and running in about six weeks, right on time. One member of the Microteam used it, but only for a while. Two and a half months after Peck finished, Firth's simulator became functional. Two months after that, Firth had refined it. He gave the Microteam a full-blown version of Eagle in a program — a wonderful machine, of paper, as it were.

If Firth had not built his paper Eagle, the lab might have looked like a crowded car on a commuter train: Hardy Boys and Microkids bumping into each other, all tempers on edge. Everyone arguing about whose turn it is to update boards. West threatening to throw them all out of the lab and do it himself. The debugging proceeding ever so slowly. Microkids writing up sections of code, then waiting in line for a chance to test them on a prototype, and after getting a crack at the prototype and finding their code doesn't work, having to set up trace after trace with logic analyzers, trying to see what's going wrong. They will be lucky if they finish this computer in three years.

As it was in fact, the Microteam could test their code right at their desks, via their own terminals. Firth's simulator was a program stored inside their computer, the Eclipse M/600 — Trixie. They merely had to feed into Trixie the microcode they wanted to test, order up the simulator, and command it to run their code. They could order the simulator to stop working at any point in a microprogram. The simulator could not tell the microcoders all by itself what was wrong with their code, but it arranged for the storage of all the necessary information about what had taken place while the code was running, and would play all of it back upon demand. Thus, without having to invent ingenious approaches with logic analyzers, the team could examine each little step in their microprograms. They could find out what was going wrong in an instant, in many cases. In the Microteam's small corner of the world, Firth's was an heroic act.

When they first moved to Westborough, in the summer of 1978, Firth and his wife, Lynn, lived in an apartment not far from Building 14A/B. His working day began in the shower, and would continue as he walked to Westborough on those summer mornings. A lone figure walking along the side of the access road — a young man, stocky and graceful, with long black hair cut straight as a hedgerow in a line just above the shoulders, wearing glasses with large black frames and often a blue Windbreaker, and black shoes. At that hour, there would be a torrent of cars — grim faces behind the windshields — hurrying toward the plant. But Firth hardly noticed the traffic. He was building his simulator. It was not a trivial task.

Firth had to write a program for every microverb. Then he had to shuffle and reshuffle these programs. In the material Eagle, all the microverbs in a single microinstruction would be executed in parallel, at more or less the same time, between two ticks of the computer's clock. But the simulator was a program, in which events occur one step at a time; it could execute the various microverbs of a given microinstruction only serially, one after the other. And the order in which the microverbs of a given microinstruction were executed was often crucial; one verb could in effect cancel another.

UINST also defined cases in which one microverb would alter

the meaning of another one and cases in which one microverb would tweak two different pieces of hardware and mean two different things to each of them. From time to time, while he was building the simulator, features of the microcode and hardware of the material Eagle would be altered; Firth had to change his simulator accordingly. Moreover, he had to make his abstract machine useful to the Microkids. How would the people who were going to use the simulator extract information and give instructions? This is one of the most important and most often neglected issues that faces any programmer. Firth began his work pondering the question, and it showed. He made his paper machine completely "interactive."

"I guess nobody considered it a possible project in the time frame," Firth said, "but I didn't know that. To me, it seemed, well, challenging. It could be done. I was able to do it because I had no idea of what was going on. Usually it was: 'We gotta have this feature in the simulator by tomorrow or poor old Jon Blau won't be able to do a thing.' That's usually how it was."

For two summers during college, Firth had worked as a manual laborer for a company that produced computerized junk mail. He'd often found himself in the back room with nothing to do, and he would scout around for software documents that had been written for the company's computers. One day he found a stack of programs that a company engineer had created. Leafing through them, he found a glaring error, one that would make it impossible for the computers to address properly junk mail bound for California. Firth had just begun to study programming, but the error was "just obvious" to him. Remembering this incident years later, Firth said that the engineer had probably been "programming by rote. He wanted to make his program look like programs he'd seen before, and that clearly wasn't gonna work." Firth always tried to avoid such an approach. "I like to work around 'why,'" he told me. "I prefer not to know the established limits and what other people think, when I start a project."

Firth said he thought his interest in electronics began when he was five years old. He was at a neighbor's house, playing with an electrical toy. The neighbor's son, somewhat older than he was, came in and warned him that he would get electrocuted if he did certain things with that toy. But Firth had already done all those things.

"I find myself very much of a loner," he said. "When I was younger, I would sit and make model boats and planes endlessly." I saw him at one of the team's parties, and he was very jolly, talking fast. He often ate lunch with some of the other new recruits in the basement and had, he felt, some fine conversations with them. He said he liked technical discussions, but he also liked to talk about other matters. "Like the ultimate meaning of life," he said, with a grin. "I like to do strange things, go off and watch a tree for an hour. I can do that. I've always found that I'm a little off the beaten track. It doesn't bother me too much. I had a big interest in abstract music in high school and no one else liked it. Maybe I'm right, maybe they are, but as far as I'm concerned they're entitled to their incorrect opinions."

Firth didn't often mingle with the rest of the group, after work. He said that was because he was married and because he had moved some distance away from Westborough. I asked him, "Do you miss the society a bit?"

"I think I've always missed it, personally," he said.

Firth, who was born in Canada, went to high school in a suburb of Chicago, "a neighborhood in which everyone was moving somewhere upward." His father was a regional sales manager for a company in the area. Firth excelled in high school, was placed in all accelerated courses, and had no trouble with any of them — and "no challenge," except the ones he found for himself in music and computers. He, too, met the machines in the form of "an ancient IBM." The high school kept its machine in a little room, which was left open for him during the summers. He would go in and program the computer all morning, and in the afternoons he'd practice with the band. It was one of those first-class midwestern high-school marching bands. Firth, in a big drum-major's hat, once led the band through a performance at Soldier Field in Chicago. He also played the contrabass clarinet, and twice performed with his high-school band under the directorship of Arthur Fiedler. He earned his spending money playing the electric bass at dances.

When Firth got to college, he felt he had to choose between music and computers. He had a flaw as a musician, he decided: he would always make a mistake. And he felt he would never be quite as good at music as he'd want to be. Programming came easily to him, although he found that he had to work extremely hard to do it very well. "I hate to say it, but in a way I feel that I can perfect it."

He was getting accolades now for his simulator. The team would be nowhere without it; everyone said so. Firth said: "I can't say that in all cases I'm extremely pleased with it. I guess, finally, it is a pretty crafty little thing, but there's some real crud at the foundation."

We were sitting at a bar, one long quiet afternoon. I asked Firth to explain his abstract machine to me. He tried. "Okay," he said, "I come to the simulator with the machine instructions to add two numbers. I also give you the microcode that tells the computer how to do the ADD. Okay. I say, 'Simulator, execute the macro-instruction, ADD, at location so-and-so in simulated memory.' So I also had to simulate memory. But I also forgot to mention that when you wrote your microcode you also coded some decode information, which is information that the IP simulator will use." He said all this and more at such a clip, one digression leading to another, that I had to laugh. He laughed, too. He agreed he was a sucker for a digression.

"You kept it all in your head?" I asked.

"It was just obvious to me," he replied.

Firth had spent about three months getting all the concepts behind his simulator straight, and he actually wrote the program in about a month. The rest of the time he spent on refinements. He could write two hundred to three hundred lines of code in his mind, but he had a hard time remembering his own phone number.

Luckily for him, somewhere in the bowels of Westborough existed a computer, connected to the building's phone system. Through it you could program into your own office phone the numbers you called most often. Forever afterward, you could dial those numbers by punching out a three-digit code. Firth could remember the three-digit codes. He had completely forgotten his own phone number. It was a great relief to be able to do so. He did keep his number on a slip of paper in his desk drawer, though — just in case he should ever need to know it.

9

A WORKSHOP

WEST USUALLY LEFT for work a little after seven in the morning and set out for home a little less than twelve hours later. The drive took only about twenty minutes, but the distance he traveled couldn't be measured that way. "Probably somewhere on 495 it all changes," he said, and it did seem, on at least several mornings, that somewhere en route toward Building 14A/B he would grow quiet and a stiffness would gather around his jaw and in his shoulders, as if he were preparing to lift something. He talked facts when he got near, if he talked at all. By the time he drove into the parking lot, he had both hands on the steering wheel. In the evening, he would go out wearing the same demeanor, but by the time he had turned onto 495 and had been cruising down the highway awhile, he might take a hand off the wheel and he could become almost loquacious. One evening, for instance, he said suddenly, while driving along, that he wished he could bring his guitar to work and play with some of the youngsters in the group who were musicians. This seemed to signify that he had crossed the frontier from the land of Data General to his other domain.

West didn't smoke cigarettes while he was at work. Away from Westborough, between sunset and bed, he might smoke a pack or more. Once he muttered that smoking wasn't harmful if you didn't do it at work. Had Cotton Mather returned as a spokesman for the tobacco industry, he might have advanced such a theory. Of course, West knew it was silly in any literal sense, and he uttered it barely loud enough to be heard. Some nights he would go away from Eagle and play music with friends and acquaintances, sometimes all night long, and then, fingers raw from his guitar strings, he would drive right in to work and become once again the tough, grim-looking manager. One evening that winter I said to him that I didn't think it was really possible to be a businessman and a dropout all at once. West said, "But I do it."

Ever since he had gone to work for Data General, West had been talking about quitting. Someday he'd wander off. In a sense, he already did so every day. West would drive away from Westborough — a place of sharp edges, functional, new — and twenty minutes later, he would arrive at his farmhouse. It sat on a country lane. A wooden plaque near the front door announced that it was erected in 1780. For some years after that, the residents must have kept busy adding things on. It was a place of many nooks and crannies. An ell went off at a right angle to the main structure; to the ell was attached a barn; and to the first barn was attached another. Behind the barns was a tall wooden silo.

West had previously owned a new and smaller house in the vicinity. When there was nothing left for him to do in the way of renovations, he had sold it and had bought the farm, with its leaking roofs and sagging, rotting barns. In several years he had almost completely restored it. He had had some tenants once, who on Sunday afternoons would set up lawn chairs outside and watch him work on the structures. West resented that invasion of his privacy, but you could understand their interest. He was the sort of carpenter who transformed things. "Tom likes to do things 'right' — and I mean quotes around 'right,' " said his wife. The evidence lay all around. He'd rebuilt roofs and walls and jacked up sagging girders, and what he had finished had all the right touches. He'd taken a room in which no corner was square, no wall plumb, and had transformed it into an airy kitchen. The cabinets looked perfect; how patiently and carefully he must have worked to fit them to that room's slanting walls and quirky corners. In the living room was a lovely mahogany coaching table with invisible hinges; West had built it from scratch. His crowning domestic achievement, though, was his basement workshop.

Most of the basement's walls were made of fieldstones, laid up dry originally, but covered now with cement in such a way that you could see the outlines of the boulders. This masonry had not been done without some communal effort apparently, for on one wall, in black paint, this question was inscribed:

What's A Place Like This Doing To A Nice Girl Like You?

There were several chambers. In one sat most of the machines: a lathe, a shaper, a radial-arm saw, a band saw, a drill press, a sanding machine, two grinding wheels (with goggles nearby), and an old belt-driven table saw in immaculate condition. There was a spacious woodworking bench with a handsome wooden vise, and, hanging above it, wood clamps, chisels, backsaws, crosscut saws, coping saws, keyhole saws — you'd touch the edges of their blades with care.

"A window on West's soul," theorized an old friend of his, speaking of this basement. Need a pencil sharpener? There was one in every chamber, right where you wanted it to be. Some music? A couple of loudspeakers sat on shelves in corners of the machinery room. A phone? There it was. A chair? One for every mood. Beer? The old refrigerator in the corner of the front chamber was fully stocked. A hammer? Practically every kind ever invented right in its place. A coat? Under the stairs on wooden pegs hung a blue denim jacket, an old slicker, a logger's jacket, a sailor's blue pea-coat — all somewhat worn and faded but giving off a scent like laundry on the line.

On posts and basement windowsills were tacked-up aphorisms, mementos and photographs. One card read: "The appreciation of pleasing decay is an important one because it is so often neglected." *It's Easy To Do It Yourself,* said another placard. A hand-lettered wooden sign read: GROUND BONE FOR SALE. A button reminded you that there had been a "Great Boston Kite Festival." A small photograph showed a robed figure astride a camel in a desert. "What's this?" a visitor asks, and West pounces: "The first known picture of Jesus Christ," he says.

Among miscellaneous items reposing here and there were an old brass cleat, fishing rods, a compressor, and a rusty something or other from the olden days of farming. And there was a room for metalworking, suitably equipped.

The storeroom lay off the machine shop. Along three of its walls, from the floor to the high ceiling, were shelves of glass jars, recycled coffee cans, cardboard boxes — all labeled in a firm clear hand, as follows: "Special Car Tools," "Gravely," "Flat Head Lifters," "Electrical Things," "Telephone Stuff," "Antenna Stuff," "Shoe Polish," "Brushes," "Good Brushes," "Rockets," and "More Rockets." I turned and there were two tall bookcases, also stocked with jars and coffee cans, containing, with all the sizes labeled, "screws," "washers," "bolts," "tacks," "rivets," "plugs," "bearings," "nails," "springs," and "shaker pegs." This was an adventurous boy's pockets. This was a craftsman's workshop. It would be a fine place to spend a Saturday morning. Everything was dusted, swept, sorted, labeled, hung near at hand and put away in the proper drawer. Could this be the workshop of the man who wrote on his Magic Marker board at work, "Not Everything Worth Doing Is Worth Doing Well"?

West often talked about building the "right" machine. He meant "right" in the commercial sense, emphatically.

But one day, back near the beginning of Eagle, Rosemarie Seale had gone into West's office and asked him, "Is it going to be a *good* machine?"

West said: "Yes, Rosemarie. It's going to be good."

As West's father had risen toward the top of AT&T, the family had moved many times. There was one year, when West was in high school and the family was residing in Lincoln, Nebraska, that West could not remember at all. He did recall that he was always working on things. He once bought a run-down little sailboat, stripped it to the bare wood, and rebuilt it, until it did gleam. As a finishing touch, he carved a set of cherry-wood belaying pins. Then he built a trailer and, behind a succession of cars whose engines he had also rebuilt, he towed his sailboat around the country — to Illinois, to Oklahoma City, to Martha's Vineyard.

He went to Amherst College, in western Massachusetts, where he studied the natural sciences. He did so without academic distinction, and it happened that Amherst was just then embracing a new Calvinist fad called the underachiever program: young men whose brains seemed much better than their grades were expelled for a year, so that they might improve their characters. At Amherst, certainly, and possibly in the entire nation, West became the first officially branded underachiever. It was something he'd always remember.

One of West's fondest memories was of playing in the town band with his father, in the little burg where they resided in Illinois. West played the trombone then, and had taken up the guitar since. Now, expelled from Amherst, he spent his year of exile in Cambridge, Massachusetts, unrepentantly playing the guitar. It was the very early sixties. West played at coffeehouses. He fell in among folksingers. He knew some of the famous ones before they became so. When he returned to Amherst the following year, he spoke of little else.

Trying to name the social change he thought he'd seen beginning, West would say, "People were leaving Harvard and becoming masons." As for himself, he decided to become an engineer. Some of his friends were astonished. The very word, *engineer,* dulled the spirit. It was something your father might be interested in.

"I think I wanted to see how complicated things happen," West said years later. "There's some notion of control, it seems to me, that you can derive in a world full of confusion if you at least understand how things get put together. Even if you can't understand every little part, how infernal machines get put together."

A classmate at Amherst remembered West as "smart — off the charts — but also naive. No, not exactly naive, but like a boy — uh, romantic. He believed in pie in the sky."

Indeed, West did not intend to do any old sort of engineering. He thought he would find a place in the space program and help to build the electronic equipment, monumental in its complexity, that would send men to the moon. On his own, near the end of college, he taught himself some digital electronics. But a few inquiries led him to feel that the interesting work in the space program was already spoken for. He managed to get a job with the Smithsonian Institution instead.

For the Smithsonian, West built and carried all over the world, to various satellite tracking stations, a series of digital clocks that told exactly the right time. At a border town in Colombia, the authorities mistook his clock for a newfangled weapon and threw him in jail. He went to sea. He saw Africa and Asia. Those were West's days as a far-wandering engineer, and when he reminisced about them, he certainly made them sound romantic. After seven years, however, he quit. He was married now and a father.

Old friends and acquaintances whom West had known back in Cambridge had become famous or semifamous musicians. He expressed no illusions about aping their success, but it made him want to take up the guitar seriously again. He'd support his family with some easy, mindless job. He figured he had to find a job that would keep him out of the military draft, for this was during the late sixties, when politicians still talked of nailing the coonskin cap to the walls of Cam Ranh Bay. A couple of the older engineers in West's team would tell the same story; they avoided the Vietnam War by joining companies that were making things for it.

Because jobs in the field appeared to be numerous and because not far from his home in central Massachusetts several companies had set themselves up in the business, West decided to become a computer engineer. At the Smithsonian he had learned to design digital circuitry — logic design, as it is called — but he in no way qualified as an engineer of computers. He had gone to school at a time when students of physics still carried slide rules and only the rare college had a computer that students could use. So he went to the local library and took out all of its modest collection of books about computers and studied them on the deck at the back of his house for about six weeks; then, when he felt he had mastered enough of the jargon to talk a good game, and in a hurry, lest he forget everything that he'd read, he talked his way into a job at RCA.

It did not work out as he planned. "I thought I'd get a really dumb job. I found out dumb jobs don't work. You come home too tired to do anything," he said. He remembered a seemingly endless succession of meetings out of which only the dullest, most cautious decisions could emerge. He remembered watching himself play with his thumbs beneath the edges of conference tables for hour and hours. Near the end of his time at RCA he got to work on projects that interested him. He saw a few patents registered in his name. He became what he'd pretended to be, a real computer engineer; but by then, RCA had lost a fortune trying to compete with IBM and was getting out of computers. The time to change jobs was upon West again.

Data General lay nearby. He went in for an interview. An executive sneered at his credentials, asked him what made him think he could build computers, and then hired him. Someone in Personnel told him to go see a certain engineering-team leader and find out what his problem was, and in no time at all West found himself working on a state-of-the-art computer, the first Eclipse.

West remembered working on the prototype of the Eclipse. Most evenings the company's president, Ed de Castro himself, would appear in the lab, eating a Fudgsicle. The soft-spoken de Castro wouldn't say much. He'd ask a few questions, ones that seemed remarkably acute to West. The questions and the man's mere presence made West feel, again and again, "This project is really important."

Then one night, without any warning, de Castro said to him softly, "Got that fucking pig working yet?"

West was startled, then amused, and finally, though he could hardly explain it to himself, aroused. *No. But I will,* he wanted to say to de Castro.

For some time during the debugging of the first Eclipse, West was ill every morning before work — a psychological form of morning sickness, perhaps. But when the job was done and he went to the factory floor and saw a long file of brand-new Eclipses come gliding down a conveyor belt, some great delight, which he would describe as "almost a chemical change," came over him, and what he wanted most of all to do then was to do it all over again someday, only better.

To the observant Rosemarie, it seemed that West was always planning. She began to believe that he planned almost everything that happened during the season of Eagle. As time went on he seemed to grow skinnier and skinnier before her eyes, as if the job and all that planning were somehow consuming his flesh. Once in a while she would look into his office. He would be staring at some paper and wouldn't notice her standing in the doorway. She would watch for a moment. "Why is he doing this?" she wondered. "He belongs in the north woods somewhere, canoeing and fishing and appreciating nature. He doesn't belong here."

Long afterward, at a time when she found herself talking about West in the past tense — "Geez, I hate to talk about him in the past" — the question of West's motivation still had importance for her. Most people, she reasoned, do jobs because they are told to and might get fired if they don't obey. But certainly West didn't have to drum up Eagle and waste away over it. Indeed, from her perspective, it really did seem as though the company didn't want the project undertaken at all. "So why is he doing it?" she asked.

"There's a big high in here somewhere for me that I don't fully understand," said West. "Some of it's a raw power-trip. . . .

"The reason why I work is because I win. . . .

"Realistically, I've got some stock in this company. I gotta help keep it afloat for a while."

Someone once suggested to West that he wanted to build Eagle out of love for chains and whips. West lay awake several nights worrying that it might be so.

"I'm sitting here burning myself up and doing it because I like it. You wouldn't have to pay me very much to do this," he said one night while he sat fretting, sick to his stomach over the slow progress of the debugging.

Later: "I'm trying to talk myself into quitting. . . ."

"Not many people around here would admit to being in business," he said. And: "What makes this all possible is doing this and putting money on the bottom line and not having to go all the way with the capitalist system. . . ."

"What makes it all possible is the kids."

When he talked about his reasons for wanting to build Eagle, West might have been trying to figure out how to get one of his lieutenants to sign up. He said he wished someone could explain his reasons to him. He said once, "De Castro knows what makes me go." He smiled. "The bastard."

He also said: "No one ever pats anybody on the back around here. If de Castro ever patted me on the back, I'd probably quit."

I traveled with West to New York. We stopped at a grocery store in which the cash registers were equipped with one of those devices that reads the price of an item automatically, a computerized checkout system. This one wasn't working well. West got down on his hands and knees and poked his head in under the cashier's counter to have a look at the thing.

The clerk made her mouth an *O*.

When West came out, dusting off his hands, he explained that he had helped design this particular model when he had worked at RCA. "It's a kludge," he said grinning.

The clerk had some trouble figuring what the beer we bought ought to cost, and as we left, West said, out of her earshot, "Ummmmh, one of the problems with machines like that. You end up making people so dumb they can't figure out how many six-packs are in a case of beer."

West didn't like digital watches particularly. Anyone who dared to consult such a chronometer and in his hearing say, "The *exact* time is. . . ," could expect to receive the full force of his scorn, for being such a fool as to think that a watch was accurate just because it had no hands.

He harbored suspicions about people who kept their own computers to play with after working on computers all day. In his office, so tidy it looked stark, West's only filing cabinet seemed to be his wastebasket, and although he did hook up a computer terminal in his office on a couple of occasions during the seasons of Eagle, he never kept it there for long. He couldn't wait to get the computer out of his sight — or so it seemed to Rasala. Rasala was amused, but he also wondered how in the world West got along without daily access to a computer.

West didn't seem to like many of the fruits of the age of the transistor. Of machines he had helped to build, he said, "If you start getting interested in the last one, then you're dead." But there was more to it. "The old things, I can't bear to look at them. They're clumsy. I can't believe we were that dumb." He spoke about the rapidity with which computers became obsolete. "You spend all this time designing one machine and it's only a hot box for two years, and it has all the useful life of a washing machine." He said, "I've seen too many machines." One winter night, at his home, while he was stirring up the logs in his fireplace, he muttered, "Computers are irrelevant."

Building them and, especially, getting them out the door still interested West, though. In the basement was an engineer named Dave Bernstein, by general consensus a brilliant circuit designer.

Though just twenty-seven, Bernstein was running the group that designed microcomputers. A veteran of FHP and EGO, labors that he never got to complete, he felt eager to finish something. So that winter Bernstein built a new microprocessor entirely by himself. He got the last bugs out of it early one Sunday morning and on the instant let out a whoop, a cry of pure joy — his triumphant howl, appropriately enough, echoing down empty hallways.

West had known such exultation himself — the joy derived from mastering machines, both building and repairing them. My wife mentioned to West one day that our son's record player had broken down. "Where is it? I'll take it with me," snapped West, adding with a look that my wife found a little frightening, "I can fix anything." What the thing was, whether a car's engine or a computer, did not matter; but since computers were among the most complex of all man-made things, they had always seemed to him, he said, to pose interesting challenges. Eagle, in this regard, was something special.

In his office one relatively serene afternoon, West said he had heard that IBM had canceled plans for a certain new computer, because the machine promised to be so complex that any given engineer would need more than a lifetime to understand it fully. "I don't know why they didn't just build the thing and see what it would do," said West. Eagle's complexity fell far short of that mark, but it was complicated enough to defy single-handed efforts. "I always wanted to do something like this," he said. "Build something larger than myself.

"Among those who chucked the established ways, including me, there's something awfully compelling about this," West said of building Eagle. "Some notion of insecurity and challenge, of where the edges are, of finding out what you can't do, all within a perfectly justifiable scenario. It's for the kind of guy who likes to climb up mountains." A couple of engineers in the group had taken up the rather dangerous hobby of rock-climbing. West may have had them in mind, but it sounded as though he was thinking of the scaling of Everest.

On the wall of West's office, beside his chair and a little above his head, were the pictures of some of those old computers, the machines that he could not bear to examine. Were they there to comfort him or to keep him nervous? "Everyone assumes that this one will work because all those others have," he said, rolling a pencil around in his thumbs. The fact of a string of previous successes, though, could imply the imminence of failure. "Realistically, you gotta lose one sometime," he said with a small smile.

I was lying in West's guest bedroom very late one night. His wife and two daughters had turned in long ago. On the perimeter of sleep, I heard West out in the living room take up his guitar and sing. He sounded rusty, but his voice, a tenor, could carry a tune nicely. He did not sing the sorts of songs that I gathered he played currently with his friends at their jam sessions, but once-popular folk songs — "The Banks of the O-hi-o" and the like. Those are seductive ballads. If you listen to them long enough, you can start believing that your way in life is strewn with possibilities.

10

THE CASE OF THE MISSING NAND GATE

LATE IN APRIL, after the first deadline had come and gone, Ed Rasala was sitting in his cubicle, examining his third, revised debugging schedule. Down the hall past his open doorway came Dave Epstein, one of the best of the Hardy Boys, whose humorous look, a grin that almost envelops his eyes, can by itself provoke smiles. Epstein was carrying a wire-wrapped board. He was holding it in both hands like a tray, with the side containing all the wires facing up. The thing looked terrible, the very picture of a kludge. For out of the nest of wires on the board, three little strands of wire stood straight up in the air, attached to nothing, with little bits of adhesive tape wrapped like tiny flags to their loose ends.

Rasala looked up, saw Epstein and his cargo, closed his eyes, shook his head, and looked again. "Hey!" Rasala cried.

Epstein stopped. He poked his head in around the partition and, grinning, made as if to offer the terrible-looking board to Rasala.

Rasala put his hands on his desk and buried his face in them.

It was just another routine day down at debugging headquarters.

In theory, it would be possible to test fully a computer like Eagle, but it would take literally forever to do so. The veterans in the Eclipse Group maintained that most computers never get completely debugged. Typically, they said, a machine gets built and sent to market and in its first year out in public a number of small, and sometimes large, defects in its design crop up and get repaired. As the years go by, the number of bugs declines, but although no flaws in a computer's design might appear for years, defects would probably remain in it — ones so small and occurring only under such peculiar circumstances that they might never show up before the machine became obsolete or simply stopped functioning because of dust in its chips. Big bugs in the logic of a machine, however, and even what might seem to be fairly small ones, have to be found and cast out in the lab. The hardware of modern computers is remarkably reliable, and needs to be. A computer like Eagle does a cycle of work in 220 billionths of a second. If it tended to fail only once every million cycles, it would be a very unreliable contraption indeed.

At first, the Eclipse Group worked on Eagle board by board, trying to make the machine functional in the most basic ways. This took months, and might have cost more time if Carl Alsing and Rasala had not finally persuaded West of the need for micro-diagnostic programs. Then they turned to the higher-level diagnostics, and the real fun began.

According to the Eclipse Group's theory of debugging a computer, you did not try to prove by exhaustive analysis that the machine was in all its details logically correct. You exercised the computer instead, and fixed it when it didn't work. The higher-level diagnostics had to provide the exhaustive analysis, in other words. They were crucial. They had to exercise Eagle strenuously. They had to be nasty, unfair and subtle, and not full of errors themselves.

A long list of programs that tested 16-bit Eclipses had been worked out over the years. The first ones on the list were fairly easy tests. They got progressively harder. Eagle would have to run all of them, and when it did they'd be able to say it was a bona fide 16-bit Eclipse. But it would also have to demonstrate that it could be a 32-bit Eagle, of course, and the diagnostics that would test Eagle's 32-bit-hood didn't exist when they started debugging. This worried Rasala. "West has everything in the company moving on. The printed-circuit boards are getting built. If there is a major flaw in Eagle, I want to find out about it now and minimize the impact." Rasala wanted 32-bit diagnostics that were tough and thorough and he wanted them right away. But the Diagnostics support group was producing them very slowly. Rasala feuded with Diagnostics. Fear made him angry. He shouted and he threatened, but none of it seemed to do much good. Those diagnostics came in gradually.

In the lab one day a sign appeared. Someone must have plucked it from the side of a road. It bore a picture of the national bird and the inscription EAGLE ASSOCIATES, REAL ESTATE. Another small device sat on top of Coke beside the microNova. This little thing had been designed as an aid to debugging Eagle, but no one ever got around to debugging the device itself. Remembering the assemble-it-yourself kits of electronic equipment that have been the starting points for many an engineering career, someone attached to the undebugged debugging device a label that said HEATHKIT. Soon these things lost their power as jokes and became part of the furniture of the lab. When they had started debugging, the wires on the backs of the boards had all been blue. They made their changes with red wires. The backs of the wire-wrapped boards got redder and redder. Slowly, painfully, Eagle was becoming an Eclipse.

Once in a while someone would find a problem, fix it temporarily, and move on. The engineer would plan to come back and make a proper repair but might forget about it, and weeks later it would cause some mysterious failure. Inevitably, this happened. They lost some time that way. Through the winter, the wire-wrapped boards held up well, but come April, constant handling was beginning to make some of the many mechanical connections unreliable. Wires and the sockets in which the chips were housed began coming loose once in a while, causing "flakey" failures. These occurred erratically and were often hard to diagnose; as the debuggers liked to say, it's hard to fix something when it's working. Here and there a bad chip impeded their progress. "Just stuff you never account for in a schedule," Rasala said. "You assume it's not gonna happen, and it always does."

In addition, the Eclipse Group's engineers were finding plenty of bugs in the logic of their design. "We went with an imperfect design," said Rasala. "We knew we were pushing it." So his schedules slipped and slipped, and slipped again. "The way to stay on schedule," he said, "is to make another one." So far, though, West could say, as he almost always did when asked about the machine, "Nothing fatal yet." Rasala would say, "It's comin', it's comin'." Usually he would add, "There's still a good chance that we've totally blown it."

For many months, no single variety of problem fed his worry, just whatever the next diagnostic program happened to turn up. But Rasala told me early on, "I believe there's always a focal point in any machine." And by May, it had identified itself as far as he was concerned. It was the part of Eagle called the Instruction Processor, the IP. Perhaps this was, at last, the bogeyman.

West had helped to debug the first Eclipse and some of the subsequent models by using an oscilloscope to look inside the machines. "An oscilloscope," said Jim Veres of the Hardy Boys, "is what cavemen used to debug fire." The Hardy Boys had much better tools, computers essentially, to debug their new computer. Rasala liked to tell them, "You guys don't know what fun it is to bring up a machine." In fact, however, you probably couldn't build a computer of Eagle's class without the help of several functioning computers, and especially not without the help of the logic analyzers.

One crucial difference between Eagle and earlier machines involved the parts of Eagle known as accelerators. These were primarily the System Cache and the IP, both of which were designed in order to eliminate the bottleneck between the machine and its storage. Think of a program as a list of assembly-language instructions and data for the instructions to apply to. A computer without accelerators runs through this list in a halting manner. It must execute an instruction, then seek out in its memory or in peripheral-storage devices the next instruction, bring it back, figure out what it requires, and finally execute it. It can take a lot more time to retrieve the next instruction and prepare it for execution than it does actually to execute it. So if you can arrange to have those first two operations performed at the same time, you can greatly increase the speed of your computer.

That, approximately, was the theory behind the IP. While it is telling the computer what instruction to execute right now, the IP is, in a sense, making assumptions about what the next several instructions in a program will be. At any give time, the IP will have one instruction in execution, one instruction "decoded" and prepared for execution, one instruction in the process of being prepared and one or more instructions already retrieved and about to be decoded. The IP has a fairly small storage compartment of its own, in which it keeps instructions that are likely to be required next and also instructions that have been used recently. Computer programs tend to repeat themselves, or to "loop." In the best case, when a program loops, the IP already has the next instructions in its storage; therefore, it doesn't have to send out for them, and time is saved. The IP is a complex device.

The other main accelerator, the Sys Cache, also makes assumptions about what the computer will be asked to do next. It contains a storage compartment considerably larger than the IP's; like the IP's, the Sys Cache's storage consists of expensive memory chips that operate with great speed. Among other roles, the System Cache tries to keep handy commonly used instructions and data, so that if the IP doesn't have the necessary information to order up the next step in a program, it can get it quickly from the System Cache. In this situation, if what's needed is in the Sys Cache, time is saved; it would take much longer for the Sys Cache to get the necessary information out of Main Memory and then pass it on to the IP, because in Main Memory the chips operate at relatively slow speed and there are a great many instructions and data to sort through.

Accelerators represent one clever, well-known means of gaining speed, and they're well worth the hardware if what's needed is a fast computer. But from the point of view of debuggers, accelerators can be infernal devices, not because of the quantities but because of the kinds of problems they create. Inconsistency between them is one of the deadly species of crocks.

To visualize Eagle's memory system, imagine a funnel. At the narrow end are the IP's storage compartment and other small compartments located in other parts of the machine. The Sys Cache is a wider storage, and the boards of Main Memory are widest, the funnel's mouth. Main Memory holds a lot of data and instructions, including an exact copy of everything in the Sys Cache and in the IP. *Exact copy* is the crucial term.

The accelerators are forever throwing out blocks of information and bringing in new ones, making their assumptions according to certain fixed and clever rules. They must do a fair amount of internal housekeeping in order to make sure that they are always consistent with Main Memory and with each other. If, for instance, a given block of instructions is residing in both the IP and Sys Cache, both must contain identical copies of that block.

But suppose the Sys Cache is changing the contents of its storage compartment, and at just the wrong moment some electronic event intervenes, in some unforeseen way. Suppose further that this electronic event causes one or the other of the accelerators to err in its internal housekeeping. And suppose this error causes the IP to contain a block of instructions that "looks" the same but is in fact slightly different from the one that the System Cache contains. When this happens, the machine is prepared to fail. There's a time bomb set inside it. The IP is going to order the rest of the machine to execute the wrong instruction, sooner or later.

Usually, it's later, and that's the rub. The machine will fail while running a diagnostic program, the debuggers will hook up their analyzers and get pictures of it failing, and nothing they see will account for the failure, because the real problem, the ticking bomb, was set somewhere far back in time, along the winding road of that diagnostic program.

This was the worst sort of problem that they were encountering that spring.

Early one morning in the middle of May, Chief Sergeant Detective of the Hardy Boys Ken Holberger turns his brown Saab down the road toward the red brick fortress of Westborough. It's a cloudy, hazy day. He notices, though, that muted sunlight comes in at the top of his windshield. Back when the debugging began, it was dark when he drove in and dark when he left. The slow advance of the morning sun across his windshield, when he makes the turn into the parking lot, has been one of the principal ways by which he has kept track of outside, planetary time. As for other events in the world at large — wars, famines, rock concerts — he is a little out of touch. When he goes home after a day on a difficult case, and sits down in his armchair and picks up the newspaper, he can't even read. He just stares at the front page.

Most of the group are working too hard now. Around this time, one of the Hardy Boys tells his wife that Data General provides its employees with alimony benefits, as well as medical insurance. And the funny part is, his wife believes him. "We're deep in the debug. Yeah, underground," says Holberger. "Burn-out city," he says.

Holberger wears a trim black beard. In his routine actions, the way he walks down a hall, the shape of his mouth when he gets ready to talk, Holberger seems to embody assertion, as West does, but more smoothly than West. Holberger has the appearance of one to whom all things come easily. You have the feeling that he couldn't look messy if he tried. You'd never see a vinyl penholder in his breast pocket, and you won't find him hanging around the basement after work to play Adventure or discuss machines. He says he doesn't spend any time thinking about what people do with computers. "A sense of the applications is somewhat missing, but it doesn't matter," he says, and he smiles slightly. "We say the ultimate goal is to build a machine to run a multiprogramming reliability test. But I understand that people who buy computers do run other programs on them, like Adventure and Star Trek and things."

Holberger joined the company three years ago and since then has risen from the status of a recruit to a position of importance; he has chief responsibility under Rasala for the details of the hardware of this major CPU. He is the right man for the job; by general consensus, he is the only member of the group with anything like a complete understanding of the new computer's hardware.

Holberger and Rasala are good friends. Rasala seems to feel something of an older brother's admiration and affection toward Holberger. "Holberger's a sharp cat," says Rasala. The faults that Rasala finds in him have nothing to do with ability, but with the fact that Holberger sometimes moves too quickly and makes careless mistakes.

And there is the problem of Holberger's style in the lab, for which Rasala is willing to accept much of the responsibility. Holberger is known as one of the tough guys in the basement. In part, this is the result of long hours and strain. Holberger says that when he can hear his mind buzzing from caffeine he tends to be his most abrupt. He also says he rarely brings frustration home; he gets it out at work, and when he's really angry, he takes it out on Rasala, because he knows that's safe. He doesn't waste time listening to people who aren't making good, relevant sense to him, just in order to be polite. If he's working on a problem with several other engineers and feels that too many are involved, he'll simply ignore what one or two of them have to say and eventually they'll get angry and go elsewhere. In local parlance, he's a "gunslinger," one who "shoots from the hip." If you can't get what you need from some manager at your level in another department, go to his boss — that's the way to get things done. He learned this style from Rasala, who learned it from West. They take the same general attitude toward their work. "It doesn't matter how hard you work on something," says Holberger. "What counts is finishing and having it work."

Like Rasala, Holberger worries sometimes about playing the tough guy. He says he often feels sorry after he's been abrupt with one or another of the Hardy Boys, and he is consciously developing tact for the lab. "Let's see, I'm a little confused here," he tends to say now, when what he means is, "You guys are all wrong."

Holberger is married but has no children yet. He says he has more than enough money right now. He has also received some company stock. Stock options, he notes, blur issues of salary; "Data General turns people into capitalists," he says. Holberger likes the local atmosphere. The jeans, West's casual dress, remind him, he says, "that we're not at IBM." He likes not having to punch a time clock. But he knows that his freedom from company clocks doesn't stem from corporate altruism. "They don't want us to know how many hours we work. If we did, they'd have to pay us a lot more.... But," Holberger says, "I don't work for money."

For the last two years, he has been involved in projects with the flavor of crisis about them. He worked on the M/600 with Rasala and went without a break into Eagle. He has been saying of late that he doesn't want to take on any more jobs like this one, but he's also been saying that he isn't sure he means it. "It's very challenging and very interesting," he says. "There's a lot of, uh, prestige, I would say. Perhaps I like some of the things I say I don't like. It's consuming. I don't know. Perhaps I don't like it. But jobs like this aren't real common. In other companies people with our experience aren't allowed to do this, I think." He wears a wry smile. "Of course, that's how Data General gets cheap labor." Holberger has noticed that there is almost no one in the basement involved in CPU design who is over thirty-five. What happens to old CPU engineers? Holberger is twenty-six now, and though not exactly on his deathbed, he is curious about what a computer engineer does "afterward." Maybe, he says, efforts like this one can only be conducted by the very young.

"Like war," I suggest.

"Yeah, really," he says, laughing.

Holberger's father was an engineer, and so are three of his four brothers. He went to Clarkson and earned a master's degree at the University of Illinois, a Jerusalem of computer engineering, back then one of the few universities in the world where a student could do research in the hardware of computers. He hung around an old IBM machine in high school, and was taking things apart virtually from his infancy. It's something he still does. Recently, when he bought a digital watch, he took it apart. Same thing with his new programmable calculator. "I usually get things back together, too."

Significant parts of Eagle are mainly of Holberger's own devising. He thrashed out the plan for the implementation of the memory management system in many long, loud sessions with Wallach. "He's a good play on Wallach," says Rasala, proudly, feeling that he himself could not have held his own with Wallach, as Holberger did. Holberger designed most of the IP in Eagle. He regrets that it isn't two years earlier, for if it were, Eagle's hardware design might represent an advance in the state of the art and not just an example of it. But Holberger feels that he and his colleagues have taken some original approaches. "There were some general ideas out there, but the actual implementation of the IP — with all due modesty — I took some vague specs and thought it up." In fact, he made the IP run faster than those specs envisioned. As for the question of whether the IP will ever work right, he isn't worried about that. The way he looks at it, the whole machine's a crossword puzzle that he and the rest of the designers thought up, and now they just have to solve it. "I'm getting quite good at it," he says. "I can track a problem back into the twilight zone quite well."

With a touch of regret, Holberger puts the spring morning behind him and makes his way briskly through the basement — still mostly empty at this hour. As expected, when he gets to the lab, he finds Jim Veres there, sitting in front of Gollum. And right away, Holberger's back into it.

Last night, per Holberger's instructions, both Coke and Gollum were left running the diagnostic program called "Eclipse 21." Some weeks before, the debuggers ran this program and the machine failed sporadically. They didn't closely examine the failure. They decided that it was probably a flakey — most likely a loose connection or "noise" — and they went on to other diagnostic programs. Now, however, the machines have successfully negotiated all of the basic Eclipse diagnostics except for Eclipse 21. So it's time to go back and clean up the problem, whatever it is — the failure that might be noise or a loose connection.

One characteristic of the diagnostic programs is extreme repetitiousness. Each test contains a number of subtests, each one of which consists of dozens of instructions — ADDs, subtracts, jumps, loops, Skips On Equal and so on. The program has the machine perform one of these subtests dozens of times, each time with different data, and then tells it to go on to the next subtest. When the last of these subtests is completed, the diagnostic program directs the machine to go back and repeat the process all over again. The entire program, including all the repetitions of the subtests, is repeated a large number of times — say, a hundred — before one so-called pass is complete.

If, during all these calisthenics, the machine fails to execute an instruction properly, the program directs it to confess the act by sending an error message out to its console and then tells it to go on with the exercise. The debuggers can come in after the machines have been running all night and find out right away, by placing an order through the console, how many passes of the program the machine has run and how many times it has failed.

Veres has already done this. He tells Holberger that Gollum ran 921 passes of Eclipse 21 last night, with only 30 failures. And Holberger makes a face.

In this context, 921 is a vast number. It means that any given instruction in the diagnostic program may have been executed millions of times. Against 921 passes, 30 failures is a very small number. It tells them the machine is failing only once in a great while — and that's bad news, because it's hard to locate the cause of a failure that crops up only once in a while. As they say, the first step in fixing something is getting it to break. The problem could just be a loose connection or noise, though. But while noise and loose connections can cause sporadic failures, they usually do so erratically, in no discernable pattern. And when Holberger asks Veres how Coke performed running Eclipse 21, Veres tells him that the story is the same as it was with Gollum: "Nine hundred and twenty-one passes, thirty failures.

"I'm still willing to call it a noise problem," says Veres. But Veres is thinking, "Either that noise is remarkably consistent or we've got a real problem in the logic somewhere."

Holberger thinks it would be nice if noise was the culprit. This bug in Eagle has the feel of an unpleasant one. So, thinking wishfully, they concoct a few theories about noise. Finally, Holberger says: "Okay. Time to fix it."

They don't need to say much to each other for a long time after that. Between Holberger and Veres there exists a kind of technical understanding that outruns the powers of speech. Most Hardy Boys share this specialist's ESP to some degree. It's a feeling that some good chess players say they share with worthy opponents, a kind of mind reading — what Holberger calls being "in sync." To a degree, all of the Hardy Boys are loners; all say they usually prefer to work by themselves. But Veres and Holberger have found that working together, they do produce results. To Veres, Holberger is "very quick," and because of his superior knowledge of Eagle's design, Holberger can often "fill in the details" for Veres.

Holberger, for his part, is impressed with Veres, and calls him "one of the stars."

Veres was given responsibility for the IP, and he designed a large part of it with Holberger's assistance. And when the debugging began, Veres quickly picked up technique and found his own style for the lab. Holberger feels that by now Veres, too, can find his way adroitly into the twilight zone of Eagle.

None of the Hardy Boys hesitates to contradict what he considers to be a wrong technical statement, no matter who utters it. Veres can be abrupt in the lab, too. He is a tall, fairly husky young man with a stern glare. You notice this sometimes when talking to him — he's looking at you and he's really listening; it makes some people nervous. His managers' confidence in him is tempered only by their feeling that he works too hard. That is how they express it.

Veres owns his own small computing system and sometimes after a long day in the lab he will go home and tinker with it. None of the old hands would dream of doing that, but some of the recruits are hobbyists. When the veterans in the group were growing up, computers were quite rare and expensive, but Veres went to school in the age when anyone with a little money and skill could make up a small personal system. Veres says that what he does at home is different enough from what he does at work to serve as recreation for him. At work he deals with hardware; when he's at home, he focuses on software — reading programming manuals and creating new software for his own computer.

Veres has no real complaints about the work; on the contrary, his only gripe is that lately his managers have been scheduling the work in the lab in such a way that he can't always get his hands on Gollum for as long as he wants. He calls computers "the ultimate toy." He says: "I like to tinker. I like to build things." In his senior year at Georgia Tech, he got interested in digital clocks. "I built four or five. Then it was computer terminals. I built one. Then I decided I oughta have a computer to hook it up to. So I got a microprocessor and then I figured it was not worth much without an operating system, so I wrote a small operating system. I did a number of all-nighters building this computer junk."

As it happened, Veres hated the first computer he dealt with. It was a big machine that many people shared, and it just spat out work; it was a cold, distant bureaucrat of a computing system, a machine to which you couldn't talk back. Soon afterward, though, he got to use a small Hewlett-Packard minicomputer; it stood alone and one could deal with it directly. "That made it friendly."

Holberger and Veres hook the probes of two logic analyzers to various parts of Gollum, and they set the analyzers so that they will snap their pictures when the machine fails. They call this "putting on a trace." They back up the program just a little ways from the point of failure; they run it, and it doesn't fail. Another clue. It suggests that they may be facing "a cache interaction problem." In a machine with accelerators, history is important; often it's some complex combination of previous operations that leads to a failure later on. So now Holberger and Veres start the diagnostic program all the way back at its beginning and go out to the cafeteria for a cup of coffee. About fifteen minutes later, when they have returned to their chairs in front of Gollum, there is a quick flash on the screens of the analyzers. The machine has failed. They have their pictures. They pull up their chairs and start studying snapshots of signals.

They are trying to figure out exactly what Gollum is doing when it fails. The pictures and the printed "listing" of the steps in the diagnostic program give them the answer.

"Okay. It's doing a JSR and Return."

In essence, the diagnostic program is telling the machine to take a short detour off the main road of the program. Gollum is supposed to "jump" away from the stream of instructions it's executing and go get a new instruction. This new instruction should direct the machine to go right back to the place where it was, before it took the jump. This small series of operations is a little hurdle, a trick question, a spot quiz, in the midst of a subtest of the diagnostic program.

Further study also tells them that the machine did in fact jump to the right instruction, and it did return to the right place; but when it got there, it executed the wrong next instruction. This tends, as they put it, to implicate the memory system, and particularly the IP and System Cache.

"Is it hitting the I-cache?" says Holberger.

That's the next question. The IP's small storage compartment is known as the I-cache, and what they want to know is whether or not the instruction that the machine is supposed to return to and execute, after its jump, is residing in the I-cache. The IP saves instructions that it has been executing recently, so if the program has called for this instruction a short time ago it will probably be in the I-cache now, at the time of the failure. They look at more pictures and from them infer that the IP is in fact "hitting" its cache. And they go on, examining with the analyzers the contents of the I-cache. They discover that it has the wrong instruction at the address where the right one should be.

The conversation that leads them to this conclusion is characterized by alarming brevity; even a skillful computer engineer from another project wouldn't be able to follow it. A rough translation may help. Imagine that Gollum's memory system is organized like a town in which every house has a mailbox. In the computer, there is a large number of mailboxes, each with its own unique address. Inside the Main Memory, there are thousands and thousands of these mailboxes. Identical copies of *some* of these mailboxes, labeled with the same address numbers and holding the same contents, are also in the System Cache. And a smaller number of mailboxes are in the I-cache. The diagnostic program has directed Gollum to jump to a particular mailbox, to a particular "address." At that address, in that mailbox, is an instruction that tells Gollum to go to another mailbox at another address. The IP looks through its cache and find that it has a mailbox with this second address. This second address is indeed the right address, but the instruction in the mailbox is the wrong one. In fact, what's there is an "error message," an instruction that causes Gollum to confess failure on the system console, which sits beside Holberger and Veres. It's a postman's nightmare.

Time dissolves in the lab on cases like this one. When Veres and Holberger look up from their analyzers, it is already two in the afternoon. In a moment, Jim Guyer comes in, puts down his motorcycle helmet, pulls up a chair, and starts asking questions.

Talking about Guyer some time before, Rasala has outdone himself in the enthusiasm of his speech, as if portraying the person he is describing: "Guyer's stubborn. Oh, is he stubborn! And he has one *amazing* flaw. Ask him a question about any problem, any problem at all, and he will get down to the most boring, low-level detail." Rasala took a breath and went on: "Guyer's a mechanic, he likes to fix things. Holberger gets an esoteric notion of an idea and then starts implementing it. Holberger gets a thrill out of making it work, but also out of inventing it. Guyer's more of a craftsman. Guyer can build it and refine it and he works for the pleasure of getting the last bug out of it. I identify with Guyer more probably than with anybody in the group. I don't think he thinks he's a computer genius either, but just a damn good engineer."

Guyer wears a brown beard that makes his face appear to rest inside an oval frame. He is much given to laughter. He abandons himself to his laugh; it's a high-pitched rapid one, the kind that makes the laugher close his eyes and shake his head. He often leaves his shirt unbuttoned down to the breastbone. A bachelor, he likes to go rock-climbing. He was one of those whom West must have had in mind when he said that computer engineering was the sort of thing that appealed to people who liked to climb up mountains. Guyer says, "I was considered wimpy in high school."

He grew up in a suburb of Boston — not one of the very fancy ones, nor one of the tough ones, but a place in which athletic ability sometimes superceded other virtues. In his high school, he says, there were "the athletes" and "the nobodys," and he was a nobody, partly because he got very good grades and partly because he had asthma. He had trouble running the mile, but, he says, "I used to surprise people in gym class."

Guyer, too, is the son of an engineer, and was a tinkerer practically from birth. "I took apart clocks and all kinds of stuff. Lawn mowers. I loved taking things apart. Loved putting them back together, too. Just to look inside and see how it works. Hands on, that's what I liked to do." He went to MIT as an undergraduate, determined to learn something and also to have a good time, and he says he accomplished both ambitions. He got mostly A's at MIT and went to Northwestern for a year of graduate school.

He had taken up with computers in high school; his school had an old IBM machine, too. He was going to study physics in college, but it bored him. He preferred engineering; he liked touching things, especially, by then, electronic ones. During one summer, he worked for a defense contractor in Boston. "I didn't especially appreciate that," he says. But the money was good and, more important, the job gave him a chance to work on something that "had never been done before," on state-of-the-art electronics. He enjoyed this aspect of it immensely. He ignored the ultimate fruits of the project; he didn't bother to get a security clearance, and so afterward he wasn't ever allowed to look at the thing he had helped to make.

Guyer has been at Data General for three years, ever since graduate school, and he likes the pace. He doesn't care a hoot, he says, about having "the president come down and shake my hand." And he doesn't think much about money, though he won't turn it down. He seems in his busyness, among the happiest of the group. Amazed at his equanimity, one of his colleagues has said, "The thing about Jim is, you can't make him mad." And he has become a favorite of both West's and Rasala's, mainly because of his attitude toward the debugging. "He started getting into everyone else's boards, and when I saw that — bang!" West has said.

It's true. At the moment Guyer appears to be much less interested in the board he helped to design, the IOC, than he is in some of the other boards, particularly the IP. Partly, that's because he knows that right now fixing the IP is more important than fixing his own board, and Guyer, who has staked several years on the Eclipse Group, feels protective toward it. "If we don't succeed, we can kiss the Eclipse Group good-bye," he says. He also likes to work on the IP because he doesn't know how it works. "To me, if I can't do it, it's more of a challenge," says Guyer.

He has spent entire nights alone in the lab, studying schematics and microcode listings in an attempt to fathom the IP. Guyer takes naturally to the night shift. He keeps microcoders' hours — which is to say, peculiar ones — and he has amazed even some of Alsing's midnight programmers. Jon Blau remembers seeing Guyer in the lab at four-thirty one morning, surrounded by several logic analyzers, all of them hooked up to Gollum. Blau was on his way home at this hour, but after a full night of work on the IP, Guyer was evidently still going strong. "I would start sniffing something around ten-thirty at night," Guyer says. "And I just *could not* let go. I didn't know how to fix the problems that I saw, but I couldn't leave until I had a picture of them."

Often, Guyer leaves around three in the morning. A few hours later, Veres comes in, and the first thing he does is to read Guyer's notes in the logbook and study the pictures Guyer has taken on the analyzers. Surprisingly often, that's all it takes. Veres knows right away what's wrong and how to fix it. Veres and Guyer make a marvelous debugging team, but only when they aren't working together.

One time, Rasala came into the lab, gestured across the room at Veres and Guyer, and, smiling, said, "The two Jims." They were sitting face to face in front of Gollum, having an argument about some fine technical point. Guyer was saying, "No, no, no, no, no!" He proceeded to describe his own theory; his hands made bowls in the air, wrapped imaginary packages, scrubbed invisible windows. Veres sat still. You could see the muscles in his jaw flexing and unflexing. When Guyer's spiel wound down, Veres began his retort in a soft voice, telling Guyer that his theory was completely preposterous. Guyer interrupted him. Veres clamped his own mouth shut. Obviously, their temperaments don't always mix. Perhaps they have too much in common. "Guyer and I are very headstrong," says Veres. "And we each want to do it our own way. We get annoyed and lose our trains of thought if it's not being done our own way." Veres also says, later on, to gales of laughter from the other Hardy Boys: "Jim and I don't debug well together. He asks too many questions that I can't answer."

Veres designed part of the IP, including some of the part known as the I-cache, and Guyer has made this piece of hardware his number-one suspect in almost every debugging case. "There was a time," Veres will say later, "when if anything went wrong, Jim would take out the I-cache. If Gollum ran out of paper, if it went up in smoke, he would take out the I-cache. . . . I didn't appreciate Jim pulling the I-cache all the time."

Taking out the I-cache means essentially disconnecting it from the rest of the machine. If the machine fails with the I-cache in place but works when the I-cache is taken out, then the problem must lie with the I-cache. That's the theory. But sometimes it's false. Taking out the I-cache makes Eagle very tolerant of faults in other pieces of hardware and in the microcode. "There was a time when everyone was claiming that the IP was wrong, and that put a lot of pressure on me," says Veres. "And from time to time I did end up proving that the problem was not on the IP." But, true, the IP and its cache have been found lying at the bottom of a number of devilish failures, and this bothers Veres. It does not matter that the IP is a much more complex piece of hardware than many others in Eagle, nor that when he helped to design it, Veres was truly a novice and pressed for time. Like everyone else in the group, he feels what Holberger calls "the peer pressure," which leads to an absolute determination not to be the one who fails. Under the circumstances, it is annoying to have the IP blamed every time something goes wrong. Veres helped to design this piece of the computer. It is, as he says, part of him now, and he doesn't like to see it get picked on unfairly.

For all of their similarities, though, Guyer and Veres like and appreciate each other, and when alone, each will brag about the other's skill. And for short stretches they can in fact work together. They are doing so now. The second shift is just coming in. They're in that transitional period when the first shift, hot on the trail of the problem and reluctant to leave it, and the second shift, eager to get into the case, mingle and work together for a time. Holberger and Veres brief Guyer on the problem. They tell him that at the point of failure the I-cache contains the wrong instruction at the right address. There are, they figure, two possible culprits, the IP or else the System Cache.

Laughing about it, the three decide to interrogate the System Cache first.

Early in the debugging someone wrote in the logbook that the Sys Cache was "working perfectly." Since then, when others have blamed failures on this board, Mike Ziegler, the man who designed it, has retorted by quoting that line. "The Sys Cache is working perfectly."

"It was just a matter of degree after that," says Holberger, who once made the same claim for the IP. "How perfect was the System Cache? Occasionally, we had to make it more perfect."

This is, of course, just another friendly way to compete: prove that the problem lies on a board that someone else designed. Although Guyer designed neither the System Cache nor the IP, he, too, has a small incentive to hope that this problem lies with the Sys Cache. He has, he says, "been beatin' pretty hard on the IP lately." He knows that Veres feels annoyed with him. So for once, Guyer doesn't even suggest that they disable the I-cache. Instead, by common consent, they start hooking their probes to the System Cache. They look at a few pictures but receive no immediate enlightenment. Tired out now, after more than ten hours in the lab, Holberger and Veres depart and Guyer is left alone with Gollum. He spends the night with it.

It is very late. As usual, Guyer is surrounded by logic analyzers, and he is peering at their screens, when suddenly he touches his mouth, wheels around, and starts flipping through one of the big books on the table. "A flash," he says.

The diagnostic programmer, in writing Eclipse 21, has assigned data and instructions to specific mailboxes in the computer's memory system. From time to time, however, the programmer has switched the game, and has written what is known as "dirty code." What this means is that the program sometimes changes the contents of a mailbox; it moves an instruction, some data, or both from one mailbox to another. Guyer has been studying addresses. When the machine fails, the target instruction — the one it's supposed to execute — is supposed to be in the mailbox with the address 21766. Earlier on, somewhere back in the diagnostic program, Gollum successfully performs the same series of operations that it later fails to execute. And what is interesting is that when it does these operations correctly, the target instruction is actually located at mailbox number 21765. This is Guyer's flash. He takes up the huge, bound logbook and records a hypothesis. The following is a rough translation:

The diagnostic program originally puts the target instruction at address 21765, and then, sometime later on, it moves the target instruction to 21766. But the IP never gets word of the change, though the System Cache does. Now, sometime after the target instruction is switched from mailbox 21765 to 21766, the program directs Gollum to execute the instruction at 21766. The IP receives this command and looks through its cache. It says to itself, in effect, "Mailbox 21766? I've got that address and there's an instruction in it. Let's run it." But in the I-cache, the target instruction is still at 21765, and mailbox number 21766 contains an error message. In short, the I-cache contains an outdated piece of memory. Why didn't it get updated along with the other parts of the memory system? Maybe, Guyer writes, the System Cache is to blame. The System Cache is supposed to know exactly what is in the I-cache. If an instruction or data gets moved to a new address, the

System Cache is supposed to tell the IP to throw away the out-dated mailbox and get the new one, the one with the target instruction in it. Somewhere back in the program, Guyer figures, the System Cache lost track of what was in the I-cache. It forgot that the IP had the target instruction in mailbox 21765, and so, when the change was made in the location of the target instruction, it never told the IP to get rid of the old, outdated mailbox. Guyer likes this hypothesis. He records it with mounting enthusiasm; and describing it later, he repossesses the feeling, speaking rapidly, gesturing with both hands. Then he stops, puts his hands on the table, and says, "Of course, it was completely wrong."

It's very late now, past midnight. Guyer has an analyzer hooked up to the bus — a transmission line, as it were — that carries signals to and from the memory system. He is tracking, somewhat at random, farther and farther back through the diagnostic program, taking pictures of addresses that get generated. Just before he knocks off, he comes across another occasion in which Gollum performs the JSR and Return and hits the target instruction successfully. In this instance, the target instruction is residing at mailbox number 21772, in the same "block" of addresses as it is located later on in the program when the computer fails. Now Guyer knows that in the course of this long diagnostic program, Gollum is being sent to the same general area of memory, many times, to find the target instruction. There's no flash this time, though. Guyer feels "sort of neutral." "It's beginning to look more complex," he is thinking, on his way out of the lab.

Veres arrives around dawn. He has made this a habit ever since he started to get the hang of debugging. Coming early these days assures him of several hours alone with Gollum.

In the back of Veres's mind still lies a small suspicion that the problem might after all be noise. And now — much to Guyer's delight, when he finds out later on — it is Veres himself who disconnects the I-cache. Then he runs the program past the point of failure, and everything works. He puts the I-cache back in and once again Gollum fails. This doesn't prove that the IP is to blame, but it does tend to eliminate noise as a suspect, once and for all, because noise tends to cause failures unpredictably, with no discernable pattern.

Veres turns to the logbook and Guyer's notes. Guyer's made some progress. He's proven that the program directs Gollum to the particular target instruction on a number of occasions. He's shown that the program changes the address of this target instruction many times, and also that the program is switching the contents of mailboxes in one fairly small region of memory. It looks to Veres like one of those problems involving a time bomb, and one way to tackle that sort of crock is to follow the clues back to the culprit. Veres sets out methodically. First, he finds out exactly where in the program the failure occurs. It's on the fourth pass, at the 158th iteration of the subtest that contains the misaddressed target instruction.

Next, Veres examines addresses rather different from the ones Guyer sampled. These addresses are called "tags." The mailboxes in the machine's memory system are organized in neighborhoods, called "blocks." Each block contains 256 mailboxes. Like each mailbox, each block has a unique address, which is a number. This number is the tag. Veres sees in his analyzer that at the moment when the machine fails, the tag of the block of mailboxes in the I-cache is 21. But the tag for what should be the corresponding block in the System Cache is 45. The two tags should be the same. Veres goes searching to find out which one is right. The answer should reveal whether the culprit is the System Cache or the IP.

Holberger has arrived by this time and has pulled up a chair. Veres, meantime, has set up the equipment for his search. He's hooked up a couple of logic analyzers in such a way that they will record the tag numbers in the System Cache and IP, both at the time of failure and in each of 256 previous ticks of the computer's clock. He's run the program all over again, up to the point of the failure, and now he's looking back, at one picture after another. This promises to be a long and tedious morning at Gollum. Holberger retreats, to work with Dave Epstein at Coke.

Nothing turns up. By the time Guyer comes in, Veres hasn't found a single new clue. Holberger holds a short conference. "We need new ideas. We're gonna defer it."

The Microkids have just delivered a new batch of rewritten microcode, and Guyer agrees to work on testing that for the evening. Veres goes home, with the two tag numbers — the 45 and the 21 — rolling around in his mind. Which one is right? It's a simple enough question, and obviously there's an answer. It's just a matter of finding a way to get it.

"I think that in the lab Veres is the best of the college kids," Rasala says. "The reason I say that — he's driven. To solving a problem. His whole chemistry, his whole environment is that he'll just go after it and after it. His attitude is that if he can't get time alone on the machine to work on a problem, he'll come in at four in the morning. So he can do it his own way. And his own way is often right."

Veres himself feels that by the time debugging begins, designs should be more nearly perfect than this one was. A bug in the logic of a design, though discovered and fixed in the lab, stands as a slight reproach to the designer. Not that Veres doesn't enjoy debugging. He just doesn't like mistakes. And as Rasala has noticed, he doesn't just work casually to find and fix a mistake, he attacks.

So it's up early the next morning for Veres. He doesn't want any interference; he wants time alone with this bug. To Veres, debugging the machine — particularly the IP, the part that he helped to design — has become "a very personal thing. . . . A computer, to people who designed it — it's part of them. You can almost feel what's wrong with it." This problem feels like a time bomb. But how to find it?

"I get quite a lot of work done in the morning while taking a shower," says Veres. "Showers are kinda boring things, all things considered." Now in the shower, before leaving for work, he conceives a new approach.

Yesterday he tried to find out whether 21 or 45 was the correct tag by looking back from the point of failure. Evidently, though, the answer lies farther back than an analyzer can look. So why not start searching from the other direction, forward instead of backward? He's done something like this before, on a similar problem. He'll run the program up to the fourth pass, and then every time Gollum performs the JSR and Return, he'll have the machine stop so he can get a picture with the analyzer and a printout on the system console. It'll take a while to go through the program that way. It'll be like searching every luggage locker at Kennedy Airport. It's a good thing he's up early, because Holberger might not stand for this. It's not the most elegant approach, but sometimes there is no elegant approach. He's willing to try this, he's decided by the time he gets to the lab.

A few hours later Holberger drives into Westborough. The sun is in his eyes this morning, and he wonders in a detached sort of way where it will be hitting his windshield when they finish this job. Debugging Eagle has the feel of a career in itself. Holberger isn't thinking about any one problem, but about all the various problems at once, as he walks into the lab. What greets him there surprises him. He shows it by smiling wryly. A great heap of paper lies on the floor, a continuous sheet of computer paper streaming out of the carriage at Gollum's system console. Stretched out, the sheet would run across the room and back again several times. You could fit a fairly detailed description of American history from the Civil War to the present on it. Veres sits in this midst of this chaos, the picture of the scholar. He's examined it all. He turns to Holberger. "I found it," he says.

At the 122nd iteration of the subtest in question, the I-cache contains the block of instructions with tag number 21. Millions of ticks of the computer's clock and thousands of instructions later, at iteration 151 of the subtest, Veres has observed the System

Cache instructing the IP to replace tag number 21 with tag number 45. The System Cache has proved its innocence. The IP, by inference, must have disobeyed the order; at iteration 158 the I-cache is caught in the act of still harboring tag number 21. "Which, I'm very sorry to say, is wrong," Veres says. The IP, *his* board, is the villain.

Some problems are easy to find and hard to fix; some are hard to find and easy to fix; some go both ways. They have seen and will continue to encounter permutations of all three. This one was hard to find. It happens to be easy, almost trivial, to repair. Now Holberger and Veres know where the failure occurs. They move fast. Seen working at such a moment, they might remind you of a couple of airline pilots, in the cockpit of their big jet, preparing for takeoff — heroes of technique, flicking switches with both hands, reading dials, and talking to the tower all at once. Veres and Holberger play the program up to iteration 151, the place where the time bomb gets set, where the Sys Cache tells the IP to get rid of the invalidated block of instructions and to bring in tag 45. They hook up analyzers and look at a number of different kinds of pictures. When, in fairly short order, the crucial picture appears on the screen of one of the analyzers, they don't need to perform any exegesis at all.

"There it is."

"Yup."

They see the IP throwing out tag number 45 and keeping the old, invalid tag 21. A few more pictures show that the IP is, quite literally, getting its signals crossed. The IP gets from the System Cache the signal to throw away tag 21, but before it can obey, the signal from the Sys Cache gets changed by another signal coming from another part of the machine. The solution lies in delaying the arrival of that second signal, so that the IP will always have time to clear out an old block before it's asked to do something else.

The solution takes the material form of a circuit called a NAND gate, which reproduces the "not and" function of Boolean algebra. The part costs eight cents, wholesale. The NAND gate produces a signal. Writing up the ECO, Holberger christens this signal "NOT YET." He's very pleased with the name. Schematics he's seen from other companies use formal, technical names for signals. The Eclipse Group, by contrast, looks for something simple that fits and if they can't come up with something appropriate they're apt to use their own names. "NOT YET" perfectly describes what this signal does. That's the Eclipse Group's way, Holberger notes. It's the general approach that West has in mind when he says, "No muss, no fuss." It's also a way — a small one, to be sure — of leaving something of yourself inside your creations.

This is fun. They install the NAND gate that produces the NOT YET signal, and Holberger writes in the logbook, "With this ECO installed, Eclipse 21 runs 10 passes." Just one more routine chore remains. They have to make certain that this change doesn't foul up something else in the machine. So they start rerunning all the other various diagnostic programs that Gollum has already passed, and everything is proceeding nicely, when all of a sudden the console starts scratching out a message.

One of the other diagnostic programs has produced a failure. "Oh, no."

"We didn't do it," says Holberger. "We didn't do it right."

He and Veres go to lunch. Holberger feels ill and picks at his food. When they return, they hook up a trace on the new failure, but without much enthusiasm. They take some pictures. It looks as if something complex is going wrong, but they can't immediately tell what, and they don't feel like opening up another long search right away. In part, their feeling stems from weariness. But there must also be something instinctive about their reluctance to dig into the new problem. For it seems they've forgotten to do something basic, and in a moment, Veres remembers.

Veres takes out the NAND gate. He runs the failing program. The machine still commits the new failure. So it's not the NAND gate that did it. Greatly relieved, smiling now, Holberger points out that they have the IP board out on the "extender." The board is hooked up to Gollum but it's sitting in a small frame of its own, outside of the main frame of the machine. This is standard debugging practice, but boards aren't made with extenders in mind, and in some cases a perfectly good board won't work while out on an extender. With nimble fingers, they put the IP back in its proper place among the other boards, and the failure no longer occurs. Now Gollum successfully negotiates all of the basic Eclipse diagnostics, including Eclipse 21.

They've reached a milestone, but one that they thought they had reached before. There's no celebration, no sitting around Gollum with their feet up on analyzers, savoring the victory, rehearsing the battle. Plenty of diagnostics stand before them. Much trickier ones, in fact.

"A feeling of accomplishment" is what Veres says he has. "But then again, there's lots more feeling of accomplishment to go."

11

SHORTER THAN
A SEASON

IN THE LAB, at the bench along the far wall, the other members of the night shift are bending low over wire-wrapped boards.

"Oh no. Not me. I did that last night," Josh Rosen has said to Ed Rasala. Now stillness has fallen over the room. The prototypes' cooling fans, the crickets of this place, drone on and on. Rosen stands all alone next to Coke. He looks rather small among the machinery. Rosen has very black hair, cut short enough to withstand a first-sergeant's inspection. His complexion, though naturally swarthy, still manages to reveal that he has been spending his days indoors. He is wearing corduroy pants, a plain cotton shirt with a button-down collar and no necktie. He often wears a sports jacket to work, even in the lab, and his black, laced shoes and the upper quadrant of a white T-shirt showing at his throat make his dress seem more conventional than a typical Hardy Boy's uniform. He could by his looks be any young age, from late adolescence to his mid-twenties; in fact he is twenty-four. Periodically, he raises one hand to his mouth and nibbles at his nails, while he works with the other.

Rosen, who designed the board called the ALU, is trying to get it to perform addition. It is not a large exaggeration to say that everything else in a computer exists in order to bring information swiftly to the ALU for manipulation; and for the ALU, adding is the mechanical equivalent of breathing. But this evening, whenever the diagnostic program has asked the ALU to add two packets of bits, the ALU has sent out a wrong answer and then performed a series of incomprehensible actions. "It goes," as Rosen likes to say, "to never-never land." At this moment, he is trying to take a picture with the logic analyzer of what's going wrong inside.

A straight white line runs horizontally across the little blue screen of the analyzer. From a cabinet in a corner, Rosen gets an object almost exactly the size and shape of a 45-rpm record and he inserts this "floppy disk" into the tall disk-drive machine that stands nearby Coke. Lights flash immediately on the disk drive. Rosen turns to the console and types a short message. At once the console starts typing by itself. A scratchy sound, which lasts just a moment, then stops. Rosen bites absentmindedly at the nails of his left hand as he leans over and reads what the console has typed. Nails still at his lips, he turns to the analyzer.

Something has happened. The straight white line that was running across the little blue screen has rearranged itself into a jagged shape, like a diagram of two teeth on one side of a zipper. Rosen is staring at the picture, his nails raised to his mouth. Slowly, still staring, he rotates his hand and takes most of his knuckles in his teeth. For a long moment, he holds this position, frozen like the image on the screen.

It might be a painting of a nightmare by Goya. Your eye is drawn from the young man's face and the hand resting in his teeth, to the jagged line on the screen, which is in fact a picture of an electronic event that took place, in infinitesimal time, just a moment ago. Though it is a common sort of picture, often seen in the lab, all of a sudden it has become dreadful. But who can say why?

From the moment they started designing the computer, engineers were dropping out. The reasons varied, from the feeling that the machine would be a kludge, to disappointment over positions in the group's pecking order. Some may have tired of the competition within the group, of what Ken Holberger called the peer pressure: "If I screw up this, then I'll be the only one, and I'm not gonna be the only one." Some may have had trouble keeping up with the others. A few did not participate in the group's social life, and some seemed to drift away from the project. Building Eagle wasn't the best of times for everyone.

Rosen came to the group in the middle of the summer of 1978 and went to work on the all-important ALU. He felt constrained to start designing right away, before he could really study the architectural spec — before, indeed, a complete spec existed. A few months later, in August, he decided that he had chosen to use the wrong sort of chips. He told Rasala he wanted to redesign the whole board. Rasala replied, "There isn't time." In effect, Rosen felt, Rasala was saying: "This'll probably work. Put a Band-Aid fix on it."

In December Rosen brought in a design that called for far more chips than it was supposed to contain. West assigned another engineer to examine Rosen's work — a necessary act from West's point of view, but a form of censure for Rosen. A while after that, Rosen underwent his review, a periodic ritual at Data General in which your boss evaluates your performance and sometimes gives you a raise. He was handed a report card less flattering than he was used to. It was the only unflattering one he had ever seen in his brief but distinguished career as a builder of computing equipment.

Rosen felt that West and Rasala were treating him unfairly. "They backed me into a corner." Certainly, they gave him a challenging task: he was to make an ALU that would perform certain kinds of arithmetic faster than VAX and yet occupy only one board. Bob Beauchamp would say later on that this just couldn't be done. In fact, West and Rasala eventually came to the same conclusion. They decided to sacrifice certain features and keep the ALU on one board. They had Rosen proceed. In any project such compromises inevitably occur, and in the end, according to Beauchamp, the ALU turned out nicely. "I think Josh did a damn good job on that board personally. There are some very slick things in that design," said Beauchamp — and he was in a position to know, because he had written some of the microcode that primarily ordered the ALU around and because, at the time he offered his opinion, he was himself designing an ALU for another computer.

When the Eclipse Group finished designing Eagle and began to debug, West told me: "Josh is doing all right. He's hangin' in there." West never said that to Rosen. But a few kind words just then could not have cured what ailed the young man.

Rosen had come over from Data General's Special Systems Division, which produces equipment for customers' special needs. "I was *the* star at Special Systems," he said. "I got all the sexy jobs. I went to the Eclipse Group and I wasn't treated like a star." Ken Holberger, about the same age as Rosen and endowed with roughly the same amount of experience, occupied that position among the Hardy Boys, if anyone did. Clearly, Rasala considered Holberger to be the team's stellar designer. Rosen, by his own account, found himself competing with Holberger. He wanted to be "the driving force" behind Eagle's hardware. He was used to being in control of entire designs. Over at Special Systems, he had often felt free to pursue pure technical excellence. Three weeks after joining the Eclipse Group, he said to himelf, "This is all wrong."

Over at Special Systems, Rosen had felt that his boss was also his friend. I remember following West into the lab one evening early in the debugging. Bound on some urgent mission, he strode briskly. Rosen turned around. He saw West coming. Grinning, he intercepted West, saying: "It does an increment now. So we know that it can add." West did not seem to hear him. He flicked out his hand as if brushing something away and strode on. "The man never said hello to me after he hired me. Literally never said hello!" Rosen cried. "I wanted to feel I was part of what was going on, but with West the only channel of communication was through Rasala, and Rasala, I felt, was a pretty narrow filter. Everything was on a need-to-know basis." Mushroom management came as a shock to him. "I felt I had no more rights than an oscilloscope."

Rosen described his own habits of thought as eclectic. He liked to think up fresh approaches, and he liked to talk about them. Something of a perfectionist, he felt offended when he was told to perform quick-and-dirty repairs. Rasala, who had the job of enforcing haste, felt that Rosen might design and redesign the ALU forever if he let him. He took seriously Rosen's casual, digressive talk about other ways the thing might be done, and he really dreaded it. "We can't think about that," he would often snap when Rosen came to him with some new idea. Rosen would walk away to contemplate Rasala's "sheer rudeness." They had almost nothing in common. One was big, gruff, athletic, and intent on getting the machine out the door on time. The other was of a more delicate sensibility, and eager to make comely designs.

Rosen had grown up near the University of Chicago. He did not care for sports. "I have a sort of very antiathletic point of view. That's the sort of thing nice Jewish boys don't do. I was captain of a softball team in college and we used to get beaten by fifty runs and be *proud of it.*" He said, "I'm not much of a team player, I guess." This was a defect in his character, he felt, but only in the strange context of the Eclipse Group. He did not hide his attitude. That made Rasala angry. He spoke gruffly to Rosen. Rosen's interest in working on others' parts of the machine, never high, declined still more.

Rosen belonged to the generation for whom computers made up part of the scenery. He himself might have had block diagrams of ALUs encoded somewhere in his genes. Like practically everyone else on the team, he started becoming an engineer at about the age of four, picking on ordinary household items such as lamps and clocks and radios. He took them apart whenever his parents weren't looking. At ten he turned to rockets. First he made them out of match heads. Then he experimented with more powerful fuels. At twelve, he got some gunpowder in the mail from a firm of dubious repute and concocted his most powerful missile. He set it off electrically, from his "blockhouse," a basement stairwell. The rocket ascended some distance and exploded, loudly. A few minutes later, he saw a police car turn into the alley behind his home. "I gotta go visit grandma in Sheboygan," he told his parents that very evening.

He went to an extremely competitive high school, the University of Chicago's Lab School. "If you had seven hundreds on your College Boards, you were at the low end of the class." In college he had no confidence until, thinking he'd like to have a stereo — and being "cheap," as he put it — he decided to learn how to make one himself, and to that end enrolled in a course in basic electronics. "I totally crushed the course," he remembered. He became a physics major. For his senior-year thesis he began to build a device called a floating-point processor; and suddenly he was getting A's in all his courses. "No one knew what I was doing. No one understood it. They thought, 'Hey, he must be something.' "

The little processor was his first concoction in computers, and he remembered it fondly, as some remember old girlfriends or certain football games. "It wasn't a bad little box at all." He felt gratitude toward that first processor, because it enabled him to graduate magna cum laude.

Rosen went on to Northwestern, to earn his master's degree in electrical engineering. He spent his summers building electronic gear. While still in school, he built a pattern recognition processor for Fermi Labs, worked on an earth station for Fairchild Space and Electronics, and designed a signal processor, also for Fairchild.

Data General recruited Rosen by promising him interesting

work, and he got it. The first truly commercial product he designed was something known as a cluster controller, a kind of
computer terminal, which he called Hydra. After it had been built
and shipped, a small microcode error cropped up in it, and
Rosen's boss sent him to California to make the repair. Out west,
Rosen was ushered into a room and saw a dozen people using the
machine that he had designed. The sight made him tremble. It
took his breath away. He felt scared. "My God!" he thought.
"Don't use one of those. Why don't you use a real terminal?" He
felt thrilled. "That's something I designed. That's my machine.
That's not Data General's. That's me.

"You don't get to see that very often," Rosen said. "But that's
the biggest satisfaction of all."

He was only twenty-two, and he had done it all — except that
he had never helped to build a commercially important, big and
brand-new computer. He expressed an interest in doing so, and
the word got back to the Eclipse Group. They recruited him;
he had fine credentials. Then, of course, everything went sour
for him. But perhaps his personal catastrophe had started
earlier. Maybe he volunteered for Eagle looking for a way out of
a malaise already upon him. He thought that was probably the
case.

Rosen said that when he first came to Data General a few years
before Eagle, the staffer at Personnel told him, "We know how
you fellows work, and we will remind you if you forget to take
your vacations." But, he said, they never did remind him. Probably, it wouldn't have mattered if they had. He went to work at
Special Systems, and in his first year there, he was assigned so
many important, challenging projects that he not only forgot to
take his vacation, he also failed to take a weekend off. What
would obtain in the Eclipse Group also held at Special Systems.
"There was no question of deadlines. You'd already missed it,
whatever it was." He worked many eighty-hour weeks — without
extra pay, of course, but that wasn't the issue. "I had a lot of con-

trol over the things I did, and the price was a lot of pressure. If I spent only a sixty-hour week, I felt intensely guilty."

He told himself that he was having the time of his life. During his second year at Special Systems, he began to remind himself of this with some regularity. "Josh," he would say to himself, "you're designing the sexy machines."

The dialogue with himself continued when he joined the Eclipse Group and began working on Eagle. "You've always revered the people who built the NOVA and the PDP-11. Now you're one of them. You're the guy you always wanted to be," he said in his mind.

"So why," he asked himself, "am I not happy?"

By the time I saw Rosen standing in front of the logic analyzer in the lab, he did not have to ask himself that question anymore. He did have some friends in the company, but few of any sort outside of work. How could he have made any? He had spent nearly half of all the hours of all the days and nights of the last three years at work.

His experience had made him feel that Data General, more than any of the other companies he'd heard about, exploited "child labor." "Kids right out of school." "It's something of a sweatshop. It's expected that you'll ruin your health for the company." But he knew it wasn't that simple. He went on: "It's sort of self-imposed. Everybody's trying to prove themselves. Eventually you burn out."

Now he thought to himself: "I have no social life. Nothing." He looked back on his career and saw that ever since adolescence he had never strayed very far from work. He needn't have taken a job every summer; his parents would have given him spending money. But that was what he had done. "I'd been doing this all my life, it seemed. In college, you know, physics majors are masochists and proud of it. Constantly pulling all-nighters in the lab or the computer centers. But you find yourself becoming really sort of narrow." When he went to some of the parties that mem-

bers of the Eclipse Group threw, he found himself and most everyone around him talking about computers. That was nothing
new, but now he also found himself thinking: "This is a party.
We're not supposed to be talking about work."

He suspected that Rasala and West were suffering from the
same disorder as he, their irritability and rudeness toward him the
symptoms of their burning out. As for himself, he had no doubt.
As the debugging continued, he felt the pressure in his stomach. It
hurt every day. This sort of work, even the occasional bad stomach, used to be fun. "Part of the fascination," he said, "is just little
boys who never grew up, playing with Erector sets. Engineers just
don't lose that, and if you do lose it, you just can't be an engineer
anymore." He went on: "When you burn out, you lose enthusiasm. I always loved computers. All of a sudden I just didn't care.
It was, all of a sudden, a job."

Rosen went on working, through the winter and spring, peering
into Coke with logic analyzers and doing a creditable job of repairing errors in his own board. But debugging Eagle had long
ago stopped being fun, and from time to time, when Rasala teased
or reprimanded him, or when he simply couldn't face that tangle
of wire and silicon anymore, he would take advantage of Carman's escape valve and go away from the basement for an afternoon or an evening. Sometimes, before he left, he would leave a
note behind, on top of the terminal in his cubicle — usually a
faintly humorous note that would nonetheless alert his friends to
what had happened in case he didn't come back this time.

With an old friend, Rosen had once visited what he called "a
very liberal arts college" in Vermont. He was strolling through
something known as "an alternative-energy farm," when a young
woman, bare to the waist, walked by. "She was," Rosen said, "a
miracle of biological engineering." He continued: "I was so
stunned that I walked into the door of a geodesic dome. Although
blood was pouring down the bridge of my nose, I was completely
oblivious to it."

Now, one day back at Data General, his weariness focused on the logic analyzer and the small catastrophes that come from trying to build a machine that operates in billionths of a second. On this occasion, he went away from the basement and left this note on his terminal:

> *I'm going to a commune in Vermont and will deal with no unit of time shorter than a season.*

12

PINBALL

SOMETIMES THEY SLOWED DOWN, usually to wait for some new part. At the onset of such pauses, Ken Holberger felt as though he had poked his head up through a trapdoor. He noticed once again that other groups in the basement had more space for offices than they. Take a look at this lab, they said to each other. It was little. It was noisy. Remember those nights back in February when some of the guys wore coats while working on Coke and Gollum? Well, nowadays, you didn't have to go outside the lab to know that summer had arrived.

One of the Hardy Boys brought to the lab a thermometer. When the temperature reached eighty-eight degrees, Holberger propped open the door to the outer hall. "So we can get the hot breeze from the hallway." On an evening visit, Carl Carman frowned to see the door opened — it was a breach of security — and he kicked it shut. Their vice president had closed it and they did not dare reopen it. But then the thermometer went right off scale. It was too much. The Hardy Boys walked out, and finally West burst from his office. "West went into forest-fire mode," said Holberger. At long last, West got Maintenance to repair the lab's cooling system. The Hardy Boys ended their wildcat strike.

The Microteam had their own computer, but they noticed that when someone was using the simulator, Trixie ran ever so slowly. We could use another computer, they said. But don't bother to ask, they told each other; after all, we had to battle West just to get Trixie.

But the Microteam was lucky compared to the Hardy Boys. "This group has built four models of Eclipses," said Holberger, "and we haven't even got *one* for our own use in the lab. We have to share Woodstock with every other group except for Bernstein's, which has two computers of its own." It got so bad during crises that Holberger had to send out emergency warning messages. This was how it worked:

Holberger runs into a problem — he needs to reprogram a PAL. To do that he needs the services of a functioning computer. He hurries to his cubicle and turns on his terminal, which is hooked up to the Eclipse called Woodstock. But a message appears on his screen saying that the program will not run, he'll have to wait — too many other engineers are using Woodstock now. But Holberger can't wait. So through his terminal he broadcasts an EMERGENCY WARNING MESSAGE. Throughout the basement, on every screen of every terminal using Woodstock at the moment, this message now appears. It says, in effect, "Shut down your terminal at once because the system is crashing." From his terminal Holberger can watch the various responses to this false alarm. Some engineers just go on working, he notes with amusement. "The cynical and jaded," he thinks. But enough other users shut down their terminals for Holberger to run the PAL program.

Holberger grinned. "I have the feeling that's the kind of behavior West approves of."

That summer, in *Mini News,* one of two in-house company papers, an article appeared bragging that Data General was spending a larger percentage of its profits on research and development than almost any other company in the industry — more, indeed, than the vast majority of American companies of every sort. Holberger saw the article. He took it into West's office. "Hey, Tom, where's it all going?" Was the lion's share going to North Carolina? Some suspected it was — but only in moments like these, the competition had receded so much.

One of the team decided he'd like to have his own business card. Others liked the idea. They felt they deserved them. After all, some other groups in Westborough dispensed personal cards to their members. So some of the Eclipse Group took the request to Carl Alsing, who agreed to ask West. They waited outside. "Sorry," said Alsing, when he came out of West's office. "Tom says no."

"How come?"

"He just says, 'No.' "

Another story made the rounds: that in turning down a suggestion that the group buy a new logic analyzer, West once said, "An analyzer costs ten thousand dollars. Overtime for engineers is free."

Dave Peck was at one point sent to Data General's semiconductor facility in California to perform what he considered to be "a near impossible job." He didn't get it done. Upon returning to Westborough, he wanted to go back to take another crack at it. He took his request to Alsing. He and Alsing went to West.

"I'm not gonna send Peck out there for another vacation on company time," said West to Alsing, in front of Peck.

Peck had an odd, some said a corny, sense of humor, and a few in the group resented it sometimes. "He's having a good time, while we're all burning up," said one of them. But others thoroughly enjoyed him. Peck had heard West use disdainfully the phrase "software mentality." Peck, who had come from Software, believed this mentality existed and that he himself possessed it. "If you'd walk into Software, there was paper all over the walls," he said. Around the time when Data General stood accused of burning down a competitor's factory, Peck remembered, a picture of de Castro appeared on a wall in Software: the bottom edge of the photograph was charred and a sign below asked,

Would You Buy A Used Car From This Man? Peck's own office was full of posters. DATA GENERAL FREE FOOD AND BEER, said one; he had picked it up at a company picnic. "I like a noisy room," he said.

Peck was slightly plump, with a thin mustache that he often stroked. Years before, when he was working in another department, he had to deal with a peer whom he considered to be incompetent and a meddler. Feeling sorely provoked, Peck one day said to this engineer, "You're an asshole." Ordered by his boss to apologize, Peck went to the man he had insulted, looking sheepish, and said, "I'm sorry you're an asshole." He never talked back to West, however. "I don't know, I guess I never thought of him as an asshole. Just mean."

But Peck got hold of a poster advertising the movie *Lord of the Rings*. The poster depicted most of the characters in cartooned form, and Peck labeled them. He made the Hobbits Microkids — "They're kinda cute, you know" — and he put his own name beside one of the heroic-looking figures; the skinny, treacherous character, Gollum, he christened "Tom West." Peck hung the poster right next to Woodstock. But afraid that West might see it, and not knowing what would happen if he did, some of the younger members of the team moved the poster to an obscure corner, out of sight.

Many of the team, especially some of the Hardy Boys, said they felt comfortable around their division's vice president, Carl Carman. He visited the lab almost every morning and evening now. He asked them about their problems. He knew all their names. But most of them, when they ran into West in the hallways, got the feeling that their immediate boss did not know nearly so much about them. "He'll look off in another direction as he walks by, and you never see him smile." When West did speak to the masses, some felt afraid.

It seemed peculiar to some of the team. Here they were laboring mightily on the most crucial project in the company, and yet they lacked equipment, comforts and all signs of recognition from their boss. Their project had number-one priority. Their vice president said so. Who among them could doubt that what they were doing was important to the company? The problem had to lie with West. "Why are some managers effective in getting resources and Tom isn't? That's my bitch," complained one member of the team.

"Sometimes even the pencil supply seems short," said another. "I can't help but think that someone wants us to run lean and mean because he thinks we work better that way. I don't know. Maybe Carman doesn't have the pull. Maybe Tom West only talks tough to people at his own level and below. Maybe he wants us to look lean and mean to impress other people above him."

When they had time on their hands to look up from the machine, some saw that they were building Eagle all by themselves, without any significant help from their leader. It was *their* project, theirs alone. West was just an office out of which came "disconnected inputs and outputs," said one Hardy Boy. He shrugged. It didn't matter. "West may be acting as a real good buffer between us and the rest of the company. Or maybe he's not doing anything."

Alsing listened, and sometimes he smiled. "When this is all over, there are gonna be thirty inventors of the Eagle machine," he predicted. "Tom's letting them believe that they invented it. It's cheaper than money."

West had put this project together, almost single-handedly. If West did nothing else, Alsing felt, Eagle would still in some ways be West's machine. But the idea that he was doing nothing else was crazy, although Alsing could understand why many on the team got that impression. Maybe they were supposed to. West had never sat down with him and Rasala and Steve Wallach and Rosemarie Seale and said: "We're gonna bury this team. They're not gonna see anything except the machine." But West had said, in a cautionary tone: "There's thirty guys out there who think it's their machine. I don't want that tampered with. It's very useful to me right now." Another time, he had remarked, wearing his crooked little grin: "Some of the kids don't have a notion that there's a company behind all of this. It could be the CIA funding this. It could be a psychological test."

The managers had sealed off the team right from the start, telling every recruit not to do so much as mention the name Eagle to anyone outside the group. Although once in a while the pencil supply did grow short, the worst administrative problems never touched the Microkids and Hardy Boys. "We *were* buried, to the point where we were almost underground," said one of the Microkids long afterward, adding, with the air of one from whose eyes scales have just fallen, that Rosemarie served them so well they never realized the full extent of what she did for them.

By the spring, Software had begun to flood the project with programmers. Hiring its own large team of new recruits, the software team was now preparing the huge, complex body of system software that had to exist in order for Eagle to become more than an interesting exercise in computer engineering. Technical relations between Software and the Eclipse Group were complex, like those between Hardy Boys and Microkids, but no one except Wallach had to deal extensively with software. West had made Wallach the lone emissary to Software, saying whenever Wallach complained, "If you don't have system software, you don't have a machine." Wallach carried messages back and forth, and sometimes invented messages of his own, he said, in order to slip some nice new feature past West. Absorbing criticism from both sides, Wallach played courier, arbitrator and referee, and enjoyed it, he said.

As the summer came on, increasing numbers of intruders were being let into the lab — diagnostic programmers and, particularly, those programmers from Software. Some Hardy Boys had grown fond of the prototypes of Eagle, as you might of a pet or a plant you've raised from a seedling. Now Rasala was telling them that they couldn't work on their machines at certain hours, because Software needed to use them. There was an explanation: the project was at a precarious stage; if Software didn't get to know and like the hardware and did not speak enthusiastically about it, the project might be ruined; the Hardy Boys were lucky that Software wanted to use the prototypes — and they had to keep Software happy. Possibly, no explanation would have sufficed for some Hardy Boys, they were so fond of their prototypes. At any rate, no full one was offered. They were left to add this insult, if they chose, to the list of West's offenses.

From time to time, Rasala and Alsing would tell some of their troops that West was acting as a buffer between them and the company bureaucrats, but the two managers didn't go into details. To do so would have violated West's unspoken orders — "an unspoken agreement," said Alsing, "that we won't pass on the garbage and the politics." They wished sometimes that the rest of the team could get a glimpse of West when some other manager dared to criticize the Eclipse Group or one of its members. West, notoriously, kept a double standard in such matters. He criticized other groups but would brook no criticism directed toward his own. He could carry this policy to absurd lengths, Rasala thought. Sometimes West would simply ignore criticism directed toward his team. Sometimes he would answer it with questions in this vein: "Are your people working sixty hour weeks?"

Alsing and Rasala went often to West's office, closed the door, and asked him why they could not give the team more space and equipment. West probably felt that a little material hardship was good for young engineers, Alsing reasoned — that an embattled feeling provoked energy, that frills would just get in the way. "Tom's also cheap," he said. That was the old-time Data General way; one myth had it that the company used to zealously conserve paperclips. As for West's unfriendliness to his troops, Alsing had heard him say more than once, "No one ever pats anyone on the back around here. That's how it works." Alsing came away convinced, however, that West had an important strategy. "We're small potatoes now, but when Eagle is real, he'll have clout and can make nonnegotiable demands for salary, space, equipment and especially future products." Rasala came away with the same idea: "Maybe it's ego. But West has some interesting notions, ahhhhhhnd, I kinda believe him. His whole notion is that he doesn't want to fight for petty wins when there's a bigger game in town."

They didn't have to name the bigger game. Everyone who had been on the team for a while knew what it was called. It didn't involve stock options. Rasala and Alsing and many of the team had long since decided that they would never see more than token rewards of a material sort. The bigger game was "pinball." West had coined the term; all the old hands used it. "You win one game, you get to play another. You win with this machine, you get to build the next." Pinball was what counted. It was the tacit promise that lay behind signing up, at least for some. Holberger felt that way. "I said, 'I will do this, I want do it. I recognize from the beginning it's gonna be a tough job. I'll have to work hard, and if we do a good job, we get to do it again.' "

West, said Alsing, was "saving his coins," to fight if necessary for the team's right to play again.

Meanwhile, out in her open workplace, located at the junction of two corridors, Rosemarie heard the young engineers complaining about their cold, mysterious leader. She began to grieve for West that summer. In her mind she said, "They don't know him at all!" She had felt he was making a mistake in being so aloof, but now she wondered. "Hey, they're all very intelligent, very creative, and every one of them is ready to take the world and give it a spin. There has to be a strong person in Tom's position. He'd be laughed out of the department if he wasn't, and no one ever laughs at Tom."

Rosemarie kept such thoughts to herself, until long after they first began. Then they just rolled out: "Maybe the guys thought that project was a gift from upstairs, but there's no doubt in my mind — it's not even worth talking about — it wouldn't have happened if he hadn't been there, and sure, there wouldn't have been the problems, but nobody would have had the chance, including myself, to do what we wanted to do.

"I know people who felt afraid of him. But he had a twinkle in his eye! You can't be scared of someone with a twinkle in his eye who allows you to do things. I can't imagine someone else running that show. Someone else might have smiled at them and said the right words, but not many bosses would have done as much as he did for them, letting them grow in their work, giving them a chance to really do something. Tom West gave the appearance of not caring for them, but he did the things that a person would do if he did care for them. Life is odd. Is it the words? Or the thing itself? I don't think those guys could've accepted a person keeping their minds down, but I don't think they know it yet. For a lot of them, you see, this was their first working experience. They never had a boss who kept them down. I have.

"He kept it inside him. He didn't go out and complain. Maybe he didn't pat people on the back, but he didn't complain either. He was very tired. He poured himself into it. I think he allowed all the complaints to be on himself. I think it was deliberate. My opinion is he wanted them to have someone to lay it all on, all their problems, so they could get rid of their frustrations and all their problems quicker and go on and do that thing that was desperately needed. I think he set himself out there as the bad guy, but bad guy is too strong. For people to grow up they need someone to lay their problems on. 'You're the daddy.' " She laughed. "Whether he did that or not deliberately, it had that effect."

Dave Keating, to whom West never said so much as hello during the project, felt with the others that West was to blame for many of their small frustrations. But months later, when he'd had a chance to take more than a cursory look around, he remarked: "We had a lot of problems getting resources, but we've still got those problems now. Maybe it wasn't West so much." Keating also wondered in retrospect whether, under the circumstances, some bad feelings weren't inevitable; and it seemed to him fortunate that the team's members had not as a rule turned on each other or on their immediate managers. "The way West was with us, it provided a one-level separation — someone far enough away to lay blame on."

On a Friday afternoon in spring, about four o'clock, seeing that there was nothing more that they could do until they got a certain new printed-circuit board, Rasala said to his Hardy Boys, "Let's get outta here." A bunch of them made off down the hallway toward the back door of the basement. There were windows that way, and sunlight in them. They hurried toward the light, trailing real laughter behind.

West sat in his office, the door open. As the laughter died away, he rubbed his nose under the bridge of his glasses and, looking up, made his small cockeyed grin. "Well, I guess I gotta find someone to design the fucking plug."

"A plug," said Rasala, when asked about that crucial piece of electrical hardware. "You gotta have one." He rubbed his chin. "But I wouldn't know how to design a plug." It was the age of specialists. West had to look elsewhere in the basement for a maker of plugs.

Would Eagle fit into a freight elevator in Europe and the Orient as well as in the U.S.A.? West thought he'd better make sure that it could. When Rasala had gone to London with a model of the M/600, to set up that machine for its European debut, he had discovered that the computer would not fit into the elevator of the building where the announcement would take place. He'd had to take the brand-new computer completely apart, in a parking lot in London, one midwinter afternoon.

What sorts of cables and connectors should they use in Eagle? Should the machine turn on with a key or a button? Good arguments for both existed. The decision mattered a great deal to the Field Service Department. West would deal with that.

There was Software to worry about, and Manufacturing — would Manufacturing agree that they could build this machine in large quantities? There was a meeting of the Product Board coming up some time in June. There, for the first time, West would formally present Eagle to de Castro. West started preparing for that day months in advance. He was proselytizing for Eagle throughout the company now.

The Eclipse Group was missing deadlines, but the whole process was moving along fairly well these days. All those gambles — on new recruits, on software — were paying off, it seemed. But just around the time when West had stopped thinking about his risky decision to use those fancy new chips called PALs, he got word that the sole supplier might be about to declare bankruptcy. Something had gone wrong, clearly. They could not get all the PALs they wanted. Never mind getting the thousands they would need to mass-produce Eagles; they scarcely had enough to repair errors in the prototypes. Rasala took to keeping a list of how many PALs they had on hand. For months, they teetered on the verge of running out. "We could lose the whole thing right there," said West.

Data General was working on the problem through other channels. In the meantime, there wasn't much that West could do about it, except worry. So he did. The younger members of the team did not know at that time how serious the problem was. No one told them, of course.

These nights, when West got home, he headed for his living room and settled into the corner of an old beige couch, always the same corner. His pants looked baggy on him now, bunched up around his waist, as if pleated. He lay back into the cushion, staring straight up at the ceiling, and sweeping his hair back with one hand, brought a cigarette up to his lips with the other. From this position he drew portraits of members of the team, ones to whom he had scarcely ever said good morning. He knew their hobbies. He knew this one's strengths, that one's weaknesses. It was startling. Sometimes he bragged about them, even Peck. "The guy's good."

You could never reproduce a project such as this one, even if you wanted to. Wasn't that obvious? West brought up his cigarette and inhaled deeply. "Ummmmmmmh. Yeah, you can repeat it."

West stubbed out his cigarette, lit another, and went back to looking at whatever it was he saw in the ceiling. "The postpartum depression on this project is gonna be phenomenal. These guys don't realize how dependent they are on that thing to create their identities. That's why we gotta get the new things in place."

Before midsummer, while Eagle remained in the lab, still flunking many diagnostics, and while the number of PALs on Rasala's list kept shrinking, West had dreamed up the vague out-lines of about half a dozen new computers and had decided, in consultation with his lieutenants, which new machines would best suit which members of the team. "Sure," said West. "What do you think I do all night while I'm staring at the ceiling? I don't just think about boats and sailing away."

13

GOING TO
THE FAIR

EACH JUNE, in order to promote the further rise of information processing, the computer industry puts on a fair, called the National Computer Conference, and known, of course, as the NCC. This year it would run in New York City for three days. The Eclipse Group was going there for one, in their very own bus. Alsing had made all the arrangements, through Carl Carman. Good old Alsing.

At roughly six in the morning, Hardy Boys and Microkids drove in one by one, to the parking lot behind Building 14A/B. The sun was up, the day already warm, and for once they did not go inside. Instead they milled around the door of their bus. They were washed and combed and dressed for town, and the smell of after-shaving lotion hung in the air. Rasala and Ken Holberger, however, after shuffling around with the crowd for a moment, slipped away. They went straight to the lab and gathered around Gollum. "It ran EMORT for seven hours last night," said Rasala. "Pretty good." It looked as though they might take up debugging stations. In a moment, though, one of the Microkids stuck his head through the doorway and said, "Hey, the bus is leaving." Rasala wrote DO NOT DISTURB on two sheets of paper and placed them on Coke's and Gollum's consoles. He hurried from the lab. Swift partings often are best.

Several didn't make the trip. Rasala tried to talk West into coming; Holberger told West he could wear a disguise and that way have a good time without any of the team knowing it. West refused. He sent along some of the refreshments, through Alsing, with instructions not to say where they came from. In an hour or so, West would be in his office, planning for the coming meeting of the Product Board, surrounded by complete silence for a change.

Most of those wearing jackets and ties took seats in the front two-thirds of the bus, while those in shirt sleeves occupied the back. Sudden bursts of laughter from the back of the coach, followed by short mysterious silences. Alsing swaying up the aisle, dispensing from a box full of soft drinks and donuts. Holberger and Jim Guyer in a technical dispute about stereo equipment. Steve Wallach holding forth on the subject of industrial recruiters, predicting there'd be headhunters roaming the floors of the conference, in search of young engineers, whom they'd take back to "hospitality suites," ply with liquor and caviar, and talk into defecting to other computer companies. The bus driver calling out over his shoulder, "Keep it down back there!" and his injunction being answered by a roar of laughter from the rear of the bus. It took you back. Doesn't everyone remember awaking to the realization that there would be no school today, that an entire summer lay ahead, that the class was going on a trip? Is there anything more gorgeous than a holiday outing in June?

In no time at all, it seemed, the bus had achieved Manhattan. Rasala glanced out the window at his hometown without expression. Wallach, who, like Rasala, had grown up in Brooklyn, spoke with the air of a proprietor of local history, about vanished Ebbets Field and the old Brooklyn Dodgers. Bob Beauchamp, who had seen New York City only once, on a trip with his high-school class from Missouri, gazed out his window all the way down from Harlem to the New York Coliseum.

Computers and the hosts of equipment that attend them filled four spacious floors. The crowd was thick and wearing badges. One badge went by that read APPLICATIONS ANALYST. Was this someone who tried to figure out what to do with all this stuff? "Hey, check this out over here," said Beauchamp. "There's bubble memory over here," said Jon Blau. One knot of Microkids and Hardy Boys headed off across the first floor. Rasala, Alsing, Holberger and Wallach went searching for DEC's booth. When they found it, they went right to the VAX 11/780 that was on display. They knelt down. Wallach had just opened up the door of smoked Plexiglas behind which stood the boards of the CPU, when a saleswoman from DEC came over and peered at their badges. " 'Data General, Engineer.' " She made a face and shooed them away. Laughing, they went over to Data General's turf.

It was at one of these fairs, eleven years before, that Data General had launched itself upon the industry, raising its placard higher than any other company's. This had apparently become a tradition, for Data General's sign was once again this year the tallest in the hall, taller even than IBM's — raised, indeed, to such a height that it threatened to become unnoticeable. Maybe its altitude was a function of increasing competition. If so, and the trend continued, someday Data General would have to set up its booth outdoors.

The company always brought something catchy to the fair. One year, Wallach remembered, Data General was introducing a microprocessor, a computer on a chip, so they hired a belly dancer, who plied her trade at the company's booth with the chip lodged in her navel. This year Dr. Gideon Ariel, a practitioner of sports medicine, was the featured attraction. Dr. Ariel had devised techniques for using a computer to improve the performance of athletes and was working with members of past Olympic teams. Data General had given him an Eclipse. An inspired act of largess, which had gotten Data General a nice mention on NBC's "Today" show. At the moment, a TV screen was playing back that bit of television footage. "Yeah, sure," whispered Alsing. "We gave him an Eclipse with only two-K of memory and no peripherals." A computer thus equipped would be quite useless, that was the joke. It probably wasn't true, but the other engineers grinned. An old-timer, Alsing liked to think of Data General putting something over on national TV.

Dr. Ariel was up on the stage now, in sneakers and perspiring, demonstrating how his Data General system could capture in detail — in pictures of a sticklike body — a jogger's movements. He was offering vivid, computerized proof that the best way to jog is on the balls of your feet. The engineers turned away.

"Oh," said Alsing. "There's *my* machine."

He went over to the C/150, a recent model of Eclipse. It was the one for which he had written some of the microcode last summer, on his microporch. He played with some of the switches on the front of the machine.

Holberger and Rasala, meantime, wandered over to the M/600. This was *their* last machine. Rasala took up a position beside the cabinet that contained the CPU, leaning against it with his arm draped over the top. He wore a faint smile. He looked as cool as a hot-rodder at the wheel, with his arm around the obligatory girlfriend.

Alsing approached. "Say. I'm a customer, and there's something I don't understand about this computer. What's this switch for?"

"It works," replied Holberger. "But it's not useful for anything."

The community of Eagle became, for the day, unglued. I looked around and none of them were in sight except for Wallach, but he was the right companion in this spot.

A grinning fellow approached, offering his hand to Wallach. For a moment I entertained the hope that this was a headhunter about to bestow hospitality upon us. But he was just the chief engineer of another minicomputer company. He and Wallach chatted for a while, and as the man was walking away, Wallach told me, too loudly for comfort: "They've got a thirty-two-bit computer. A 1963 design."

You could have sensed that Wallach knew computers just from the way he carried himself. Approaching a display, he looked as though he were sniffing the air. A snort signified a kludge; a shrug, good but ordinary work; a judicious nod, rarely bestowed, that here at last was evidence that some people knew what they were doing. IBM required a lingering inspection. You realized first of all that IBM's large and many-sided booth was not made of ersatz materials but real oak. "You'll notice," Wallach added, "they have just the right number of black people." Women, too.

IBM seemed to be making no attempt at flashiness. They had no catchy display, just men and women in white shirts describing the machines on exhibit. These computers were mainly members of IBM's new 4300 line, which the company had announced not many months before. In fact, at the moment, IBM didn't have to talk people into buying this new kind of machine. IBM's problem just then was how to produce enough of these computers to meet the orders rolling in. At the moment, an estimated three years' worth of back orders had piled up, a piece of good luck for the many companies that, like pilot fish, follow IBM around —the so-called plug-compatible manufacturers, which produce processors and various other equipment that can be hooked up without modification to computing systems built mainly around IBM computers. In some sense, all companies in the computer industry are orchids on IBM's tree. Everyone has to consider IBM's pricing and strive for compatibility with IBM equipment.

It seemed curious: a company suffering from too much demand. Not that in the long run IBM was going to suffer more than discomfort from this backlog, but the collapse of many small, promising computer companies had begun with a similar problem. They'd announce a new product and then for one reason or another they wouldn't be able to produce it in sufficient quantities to meet their obligations. They'd asphyxiate on their own success.

But a small company had to court disaster. It had to grow like a weed just to survive.

From IBM Wallach led me to some of its traditional competitors. First, to Sperry Univac, the descendent of the first real computer company, which might have become the IBM of the industry had it not blown its early lead; Sperry had a big new machine and a slide show all about it. Burroughs had erected a small theater, with chairs set up to face a rank of computers in big white cases that looked like nothing so much as dishwashers and refrigerators; a recording of trumpets playing fanfares ushered you into this homely spectacle. The main thing memorable at National Cash Register was the pair of golden-haired women, identical twins, posing at its booth. "The bipolar blondes," said Jon Blau, when we ran into him later. It was a fairly witty remark, drawn from the language of semiconductor technology, but one of those that loses flavor during the time it takes to explain it.

Wallach took a look at all the 32-bit superminis at the fair, the machines with which Eagle, if it ever got out the door, would presumably compete. These machines were hot items now. Some estimates had it that total sales and back orders for DEC's VAX were approaching a thousand, which would translate into something like half a billion dollars in revenue, just for that one new machine.

A certain broad division between companies was apparent. The established and successful, with many wares to show, had sectioned off pieces of the central floors, setting up theaters and playhouses. The smaller, newer, less diversified had small booths along the walls. They were the little jewelry and camera stores west of Broadway. Wallach gave them a window-shopper's tour. Many of these little outfits were selling "pin-compatible memories," ones that could be added to DEC's and Data General's machines more cheaply than the extra boards offered by the big manufacturers. "We're suing some of these guys," said Wallach.

We glanced at many systems for producing graphics, one of the latest segments of the industry to experience a boom. You can draw pictures with these machines. Some of the most popular programs for graphics systems created charts of the type dear to executives with ambitious plans they want to sell to their bosses — the kind of charts that carve some commodity such as revenues into various parts, as you might divide a pie. Computers were cranking out many sorts of pie charts: multicolored pie charts, pie charts that rotated, pie charts in three dimensions. We wandered past array processors performing tricks in FORTRAN, past tape drives, Winchester disks, printers and consoles in sleek-looking packages, and the latest models for hooking up computers to telephones.

"I notice more Japanese here than in other years," said Wallach. This was a big issue. Already the Japanese had made substantial inroads into the American market for integrated circuits; and you knew that segment of the industry was worried, because they were clamoring for protective tariffs. Data General, on the other hand, had bought half of a Japanese minicomputer company — no flies on Data General.

We wandered on, casting passing glances at trade magazines, which were selling advertising and subscriptions; past software houses, which specialize in creating those crucial, ever more expensive user programs that had set the stage for Eagle; past systems houses and OEMs peddling turnkey systems — buy from us and you won't have to worry; just turn the key and your computer system works. A great deal of the overall industry, it seemed, consisted of computer companies taking in each other's wash.

Here and there we saw products of the future, such as bubble memory. After a while, however, most of what was on display began to look the same. From every booth came the insistent clamor, that here, at last, was the piece of computing gear that was the fastest, the most reliable, the easiest to use, the "most intelligent," simply the greatest in the universe. The spectacle began to overwhelm me — not so much the variety and versatility of machines themselves, but the volume of gear, the size of the crowd, the sheer number of firms.

Wallach said that two of the best-named companies were not on hand: Itty Bitty Machines (another "IBM") and Parasitic Engineering. But I saw many other names, passing by. Among others, I saw Centronics, Nortronics, Key Tronic, Tektronix and also General Robotics. There were Northern Telecom and Infoton and Centurion, which had a fellow dressed as a Roman soldier standing by its booth. There were Colorgraphics and Summagraphics; Altergo and C. Itoh; and Ball.

"Hey, wait a minute. What's Ball doing here? Aren't they the mason jar people?"

"Yeah, but they also make disk drives."

Also: the Society for Computer Simulation, and Randomex, and Edge Technology, and Van San, which sold "Quietizers." There were Datum, Data Pro and Data I/O, Tri Data, Epic Data, Facit Data, Control Data, Decision Data, Data General and Data Specialties. And we didn't have time even to glance at the wares of Itek, Pertec, Mostek, Wavetek, Intertek, Ramtek . . . Ah, Ramtek.

"In 'seventy-three," said Wallach, "there were two floors, and now there are four floors and it's just as crowded."

Norbert Weiner coined the term *cybernetics* in order to describe the study of "control and communication in the animal and the machine." In 1947 he wrote that because of the development of the "ultra-rapid computing machine, . . . the average human being of mediocre attainments or less" might end up having "nothing to sell that is worth anyone's money to buy." Although Weiner clearly intended this as a plea for humane control over the development and application of computers, many people who have written about these machines' effects on society have quoted Weiner's statement as though it were a claim of fact; and some, particularly the computer's boosters, have held the remark up to ridicule — "See, it hasn't happened."

Since Weiner, practically every kind of commentator on modern society, from cartoonists to academic sociologists, has taken a crack at the sociology of computers. A general feeling has held throughout: that these machines constitute something special, set apart from all the others that have come before. Maybe it has been a kind of chronocentrism, a conviction that the new machines of your own age must rank as the most stupendous or the scariest ever; but whatever the source, computers have acquired great mystique. Almost every commentator has assured the public that the computer is bringing on a revolution. By the 1970s it should have been clear that *revolution* was the wrong word. And it should not have been surprising to anyone that in many cases the technology had served as a prop to the status quo. The enchantment seemed enduring, nevertheless. So did many of the old arguments.

"Artificial intelligence" had always made for the liveliest of debates. Maybe the name itself was preposterous and its pursuit, in any case, something that people shouldn't undertake. Maybe in promoting the metaphorical relationship between people and machines, cybernetics tended to cheapen and corrupt human perceptions of human intelligence. Or perhaps this science promised to advance the intelligence of people as well as of machines and to imbue the species with a new, exciting power.

"Silicon-based life would have a lot of advantages over carbon-based life," a young engineer told me once. He said he believed in a time when the machines would "take over." He snapped his fingers and said, "Just like that." He seemed immensely pleased with that thought. To me, though, the prospects for truly intelligent computers looked comfortably dim.

To some the crucial issue was privacy. In theory, computers should be able to manage, more efficiently than people, huge amounts of a society's information. In the sixties there was proposed a "National Data Bank," which would, theoretically, improve the government's efficiency by allowing agencies to share information. The fact that such a system could be abused did not mean it would be, proponents said; it could be constructed in such a way as to guarantee benign use. Nonsense, said opponents, who managed to block the proposal; no matter what the intent or the safeguards, the existence of such a system would inevitably lead toward the creation of a police state.

Claims and counterclaims about the likely effects of computers on work in America had also abounded since Weiner. Would the machines put enormous numbers of people out of work? Or would they actually increase levels of employment? By the late seventies, it appeared, they had done neither. Well, then, maybe computers would eventually take over hateful and dangerous jobs and in general free people from drudgery, as boosters like to say. Some anecdotal evidence suggested, though, that they might be used extensively to increase the reach of top managers crazed for efficiency and thus would serve as tools to destroy the last vestiges of pleasant, interesting work.

Dozens of other points of argument existed. Were computers making nuclear war more or less likely? Had the society's vulnerability to accident and sabotage increased or decreased, now that computers had been woven inextricably into the management of virtually every enterprise in America?

Wallach and I retreated from the fair, to a café some distance from the Coliseum. Sitting there, observing the more familiar chaos of a New York City street, I was struck by how unnoticeable the computer revolution was. You leave a bazaar like the NCC expecting to find that your perceptions of the world outside will have been altered, but there was nothing commensurate in sight — no cyborgs, half machine, half protoplasm, tripping down the street; no armies of unemployed, carrying placards denouncing the computer; no TV cameras watching us — as a rule, you still had to seek out that experience by going to such places as Data General's parking lot. Computers were everywhere, of course — in the café's beeping cash registers and the microwave oven and the jukebox, in the traffic lights, under the hoods of the honking cars snarled out there on the street (despite those traffic lights), in the airplanes overhead — but the visible differences somehow seemed insignificant.

Computers had become less noticeable as they had become smaller, more reliable, more efficient, and more numerous. Surely this happened by design. Obviously, to sell the devices far and wide, manufacturers had to strive to make them easy to use and, wherever possible, invisible. Were computers a profound, unseen hand?

In *The Coming of Post-Industrial Society,* Daniel Bell asserted that new machines introduced in the nineteenth century, such as the railroad train, made larger changes in "the lives of individuals" than computers have. Tom West liked to say: "Let's talk about bulldozers. Bulldozers have had a hell of a lot bigger effect on people's lives." The latter half of the twentieth century, some say, has witnessed an increase in social scale — in the size of organizations, for instance. Computers probably did not create the growth of conglomerates and multinational corporations, but they certainly have abetted it. They make fine tools for the centralization of power, if that's what those who buy them want to do with them. They are handy greed-extenders. Computers performing tasks as prosaic as the calculating of payrolls greatly extend the reach of managers in high positions; managers on top can be in command of such aspects of their businesses to a degree they simply could not be before computers.

Obviously, computers have made differences. They have fostered the development of spaceships — as well as a great increase in junk mail. The computer boom has brought the marvelous but expensive diagnostic device known as the CAT scanner, as well as a host of other medical equipment; it has given rise to machines that play good but rather boring chess, and also, on a larger game board, to a proliferation of remote-controlled weapons in the arsenals of nations. Computers have changed ideas about waging war and about pursuing science, too. It is hard to see how contemporary geophysics or meteorology or plasma physics can advance very far without them now. Computers have changed the nature of research in mathematics, though not every mathematician would say it is for the better. And computers have become a part of the ordinary conduct of businesses of all sorts. They really help, in some cases.

Not always, though. One student of the field has estimated that about forty percent of commercial applications of computers have proved uneconomical, in the sense that the job the computer was bought to perform winds up costing more to do after the computer's arrival than it did before. Most computer companies have boasted that they aren't just selling machines, they're selling *productivity*. ("We're not in competition with each other," said a PR man. "We're in competition with labor.") But that clearly isn't always true. Sometimes they're selling paper-producers that require new legions of workers to push that paper around.

Coming from the fair, it seemed to me that computers have been used in ways that are salutary, in ways that are dangerous, banal and cruel, and in ways that seem harmless if a little silly. But what fun making them can be!

A reporter who had covered the computer industry for years tried to sum up for me the bad feelings he had acquired on his beat. "Everything is quantified," he said. "Whether it's the technology or the way people use it, it has an insidious ability to reduce things to less than human dimensions." Which is it, though: the technology or the way people use it? Who controls this technology? Can it be controlled?

Jacques Ellul, throwing up his hands, wrote that technology operates by its own terrible laws, alterable by no human action except complete abandonment of technique. More sensible, I think, Norbert Weiner, prophesied that the computer would offer "unbounded possibilities for good and for evil," and he advanced, faintly, the hope that the contributors to this new science would nudge it in a humane direction. But he also invoked the fear that its development would fall "into the hands of the most irresponsible and venal of our engineers." One of the best surveys of the studies of the effects of computers ends with an appeal to the "computer professionals" that they exercise virtue and restraint.

When the team came trooping back onto the bus — laughing, teasing, their faces glowing in the evening — it became obvious that many of them had not stayed long at the fair. They had gone off instead to see Greenwich Village and Times Square, and they looked refreshed. The conference didn't really fit their interests. "I don't care how computers get sold. I just build 'em," said one of the kids. Alsing remarked, "I don't even know how much an Eclipse M/600 costs."

Some in the community of Eagle cheerfully professed ignorance about, and little interest in, the ultimate uses to which the machines they built were put. But they did not hold consistently to that attitude. Some had seen in use machines that they had made, and they confessed that was a thrill. Some didn't hold to the attitude at all. "No, I think about it a lot," Chuck Holland told me. "Initially, when I was starting out, I could have worked for a company that makes machines directly for the military, and for more money. But I'm not gonna design anything that directly bombs someone."

Young computer engineers who professed anxieties about the fruits of their labors — those to whom I talked, anyway — usually named as the source of their worry the military applications of computers. One young man told me that if his company should ever start building devices of destruction, he would try to talk the executives into stopping, and if worse came to worst, he would make sure that those devices never worked. No doubt he meant it, but I think he overestimated his power. He worked for another outfit; no member of the Eclipse Group ever talked of sabotage. It would, in any case, have been an idle threat.

When the Eagle was done, according to the plan, Data General would ship the blueprints to an OEM, which would build a "ruggedized" version of the machine. Thus equipped for battle, Eagle would be sold to other OEMs, and they in turn would put it together in packages of military design for sale to the Department of Defense. Not everyone in the group believed that such applications would constitute an unworthy end, while others preferred not to think about that side of the process. What, in any case, could a dissident have done in the situation? Make sure that the plans for Eagle got fouled up and that the machine never came to life? Eagle, as it was planned, would live up to its designation as a general-purpose computer, competent at scientific and commercial chores, as well as military ones. If it did any, it could do all. A dissenter could indeed refuse to do work that might end up in the hands of soldiers: that would mean, in effect, not being a computer engineer.

Shortly after he arrived at the company, at a party for newcomers known familiarly as "cocktails with the Captain," Jon Blau had asked de Castro if Data General did business with the Republic of South Africa. De Castro replied, Blau remembered, that the company did not at that time, but that Data General wasn't necessarily guided by politics — which was perhaps an understatement. Blau worried more openly than anyone else in the group about how people used computers, but he felt confused about the subject. "A friend asked me what is my social role. Obviously, what I'm doing is in demand, but it's hard for me to pinpoint why. . . . But for every bad use of a computer, I can think of a good one."

The computer's reputation for awesome obscurity and the real complexities of the engineers' trade made barriers that were hard to cross. Their wives, some of the team said — and some of the wives agreed — didn't know much about what they did all day. "No one understands what we do," said Alsing. These youngsters led, in this respect, monastic lives. What were computers doing, going to do to society? They weren't often asked.

A number of the engineers read prodigious amounts of science fiction. Dave Keating, for instance, read three or four novels of this sort a week. "For one," he explained, "science fiction has a lot of optimism in it. I mean, we made it that far at least. Part of it, too, we can identify with the technology; and I like the imagination in those books." Several of them, at idle moments, liked to conjure up stories of their own. Chuck Holland told me: "I do a lot of thinkin' about what my mind must do. The way I think it's gonna be is that the computer will grow up with the child. When you're born, you'll be given a computer. There'll be this little thing" — he grabbed his shoulder to show where it would sit — "that goes around with you. You'll teach it how to talk after you learn how." He imagined a teenaged boy teaching his computer how to drive, telling it, "Okay, now you give it a try," and correcting it when it made a mistake. "It's gonna be an extension of me," Holland said. But maybe, he went on, a time would come when the home would be "practically a simulator," and the computer would run every aspect of a person's life. "Then we get tired of it. We start growing plants or something. Maybe slowly we will turn around and go away from it. If computers take something away from us, we'll take it back. Probably a lot of people will get screwed before that happens."

Alsing said: "I have a great idea for a science fiction story. What happens when infinite computer memory becomes infinitely cheap?" But that was on the ride back, and no one at that moment seemed interested in such speculations. They were having fun. The Microteam gave Holberger an Honorary Microcoder's Award — with some reservations, on account of a change he had made in UINST after they had already taken their unanimous vote (all such awards required unanimous approval, Alsing pointed out). Holberger accepted with a graceful speech from the back of the bus.

Someone asked: "Hey, where's the Michelob? There was a whole case of it somewhere."

"We drank it all," someone else confidently and proudly asserted.

A little later on, a case of Michelob was handed back from seat to seat, while Alsing stood in the aisle holding forth hilariously on the perils that New York City's massage parlors held for young men from the country.

They were free for a little while on a summer's night — comfortable, of course, in the assurance that interesting work awaited them tomorrow, but for the moment unbound from their machine.

14

THE LAST CRUNCH

In August, Carl Carman asked Ed Rasala when he thought they would finish the debugging.

Rasala looked squarely at his division's vice president and said, *"I don't know."*

West was greatly amused.

As the last two items on his many debugging schedules, Rasala listed the Eclipse and the Eagle Multiprogramming Reliability Tests, the hardest of all the diagnostic programs. When Eagle could play those for a full night without failing, then it would be ready to become a computer. The debuggers would, Rasala planned, run Adventure, and then they would send one of the prototypes down the hallway to the Software Department. They would "ship it to Software," and the thirty or so software engineers now working on the project would continue to endow it with the complex set of programs known as an operating system.

Each time they had approached the deadline for this consummation, West had named a new deadline, which he would announce to superiors and various interested departments. Accordingly, Rasala would draw up a new debugging schedule. West said April, and Rasala worked up a plan that might get them there by then. It didn't, so West proclaimed May to be the date, and Rasala followed suit. Then it was June, and finally West settled on the end of September. By then, Rasala couldn't stand it anymore. He wasn't making any more promises. On the other hand, he felt that they had better do it by the end of September. "Our credibility, I think, is running out."

As a group, they often got ahead of themselves. They began to celebrate that summer. As their first favorite pretext for a party, they used the presentation of the Honorary Microcoder Awards that Alsing and the Microteam had instituted. Not to be outdone, the Hardy Boys cooked up the PAL Awards, the first of which they gave to the Eclipse Group's own "CPU," after work, at a local establishment called the Cain Ridge Saloon. The citation read as follows.

HONORARY PAL AWARD
Rosemarie Seale

In recognition of unsolicited contributions to the advancement of Eclipse hardware above and beyond the normal call of duty, we hereby convey unto you our thanks and congratulations on achieving this "high" honor.

It came in a frame, and with a plastic cover. Glued to the center of the citation was a socket, such as is used to hold a PAL chip on a printed-circuit board. This being a PAL Award, the socket was empty.

At a party at his home, Chuck Holland handed out his own special awards to each member of the Microteam, the Under Extraordinary Pressure Awards. They looked like diplomas. There was one for Neal Firth, "who gave us a computer before the hardware guys did," and one to Betty Shanahan, "for putting up with a bunch of creepy guys."

Having dispensed Honorary Microcoder Awards to almost every possible candidate, the Microteam instituted the All-Nighter Award. The first of these to Jim Guyer, the citation ingeniously inserted under the clear plastic coating of an insulated coffee cup.

Somewhat later, at a time more appropriate — at the Appreciation Dinner that Carman, with coaxing from Alsing, threw for the team — the wives got their due, the Eagle Awards. Alsing wrote all the citations. Penny Rasala's read:

EAGLE
(Eclipse Appreciation and Gratitude for
Lonely Evenings Award)

This is to certify that Penny endured many lonely hours in the absence of Edward Rasala while Gallifrey Eagle was being developed, and on many occasions wondered what was the advantage of Ed being boss if he had to work 2 shifts.

It was signed, in type, by one "Gallifrey Eagle," and witnessed by Alsing.

Thus — as Alsing, himself the main instigator and organizer of parties, put it — "We congratulated ourselves on finishing Eagle and then we went back and finished it."

It was an evening in August. They still had a long way to go. But, as if in concert with the season, the Hardy Boys were stuck in the deepest lull of their campaign. They had gone back to working single shifts. Even the pace of Rasala's conversation had slackened. Walking toward the lab, he spoke in short sentences and paused in between them. "The momentum has slowed down dramatically. . . . Back in the early days, when nothing worked, it was easy to find things to do. . . . Now almost everything works. . . . The problems are harder to find. . . . They take days. . . . The people are more tired, I think. . . . The problems may be less interesting. . . . Some are more complicated. . . . This is the grind part, I think."

Only a few were working in the lab that evening. None was too

busy to talk. Seated in front of Gollum, Jim Veres discussed the
picture on the blue screen of his analyzer. Hundreds of white dots
blinked on and off while many white lines danced among them.
The screen looked like the ceiling of a planetarium, the stars
moving, white lines between them sketching out the constella-
tions. It was lovely. Veres called this picture a map. "After staring
at the map a long time," he said, "you get used to what it should
look like." By now he and the other debuggers could tell which of
the dozens of diagnostic programs was running just by looking at
the map. Did I notice how the patterns kept changing in rhythm?
he asked. If on some part of the screen images froze, that meant
trouble inside. "This picture looks healthy."

Rasala stopped in at Jon Blau's station, in front of Coke. Blau is
dark-haired with the smooth, clear skin of a child; but large black
circles under his eyes stood out in sharp contrast to his white skin.
He looked like a raccoon. Rasala sat down and visited with Blau
for a while.

Blau had become a Hardy Boy after Josh Rosen quit. Blau had
been the logical choice to take over the debugging of the ALU,
because he had written much of the microcode for the arithmeti-
cal instructions. It had been hard for him, though. "I was terribly
excited by it, then very frightened," he said. "I knew what was
going on with the microcode, but with the hardware, everyone is
ten miles ahead of me, so I've got to run as hard as I can to catch
up. Plus, I've got to keep pace. I'm keeping a note pad because
people ask me questions and I think 'I should know the answer to
that.'"

Rasala was sitting cowboy-style, his chair reversed, his arms
resting on its back. He was facing Blau and as Blau talked, Rasala
seemed to study his face.

Blau had presented to Rasala a problem of management at least
as severe as Rosen had. No sooner did Blau start work on the
ALU than he became so frightened that he had to take a week off.
Blau also liked to work, said Rasala, "from one P.M. to whenever."
He kept what Rasala called "good Alsing hours," and this habit

created scheduling problems. They had four prototypes now, but many people needed to use them. Scheduling "machine time" was tricky, and an engineer keeping quirky hours could disrupt everything. Moreover, the bugs now tended to be synergistic; someone working on a problem in the IP might need to have the ALU's expert around.

And Blau was at least as different from Rasala as Rosen; Blau was known as the introspective one, the philosopher of the group. He didn't fit Rasala's definition of a Hardy Boy exactly. "He's not as tough as the rest of us are," said Rasala. "He's a sensitive guy. He's a good guy. So if someone comes into his office, he'll tend to let them talk for hours, no matter how dumb their questions are." Nevertheless, Rasala did not speak brusquely to Blau or tease him roughly. They had what Rasala called "heart-to-hearts":

"Well, here's my problem with you, Jon."

"Well, Ed, here's my problem with you."

If you saw them talking together, you might have thought that some of Blau had rubbed off on Rasala. Or vice versa. Or both.

When Rosen left, Rasala went around for weeks saying, "I failed," and "If only I'd had more time . . . a guy as smart as Josh." No manager wants to lose people; it doesn't look good. But Rasala's reaction had seemed to border on grief. He had read self-help manuals and shaved off his beard, and he looked less imposing without it. Nowadays, it seemed, he spoke wistfully on such topics as his bouts with diagnostics and his fear that there would be no stock options for "the kids."

One day around this time Holberger came storming into Rasala's office, complaining about Jim Guyer. Guyer, it seemed, refused to stop working on a certain problem of low priority. Holberger and Guyer were like brothers, always arguing. "I'm gonna get Guyer off the machine," said Holberger.

Rasala just listened calmly and nodded and nodded. Finally, Holberger said: "Ahhh, let Guyer do what he wants. It doesn't really matter."

Rasala smiled and kept on nodding. It was catching perhaps.

Now, in the lab that evening in August, Rasala asked Blau what he was working on. They talked about the particular problem for a while. Rasala turned to me and said, "It's the kind of problem where you push in the clutch and the horn works."

"Are you trying to clean up the last crock?" I asked Blau.

"The *next* one," he said. "I used to think I was getting close."

Rasala was resting his chin on the back of his chair now. "It's hard to tell when you're finished," he said, without moving.

One of the engineers working on the new prototype named Gallifrey called out to another, who was on his way out of the lab, "Don't burn that PAL yet. I have to ECO it." And then it was quiet again, except for the hum of the fans in the machines.

"Well," I said to Blau. "I'll stop asking questions and let you go home. You look tired."

"It's a long-term tiredness," said Rasala.

"Going home won't solve it," said Blau.

For weeks afterward, Rasala repeated that thought again and again. "It's a tiredness going home won't solve." It fit his mood perfectly. The crew's wives — "and/or girlfriends," said Rasala for a joke — were growing weary themselves of the long hours. He felt under pressure from his crew to stick to one shift a day and to more or less normal working hours. Meantime, the pressure to finish by the end of September was building.

Once, Alsing had tried to guess the plot ahead of time. Months ago, he had said: "Near the end, Tom's going to declare an imminent disaster and go into the lab himself. Sometime near the end we'll do that."

A computer of Eagle's class must perform at high speed two special kinds of arithmetic called single-precision floating point and double-precision floating point. Scientific users, especially, care about this matter. Maybe more important, this is one of the few areas in which the quality of a computer can be measured and given a number, through a standard test called the Whetstone Benchmark. It is not the only or necessarily the most important standard by which to judge a computer — if you want to do trigonometry on your computer it's very important — but it is a popular test, partly thanks to Data General. For its time, the Eclipse had fine Whetstones, and Data General did not fail to brag effectively about them. West had hoped at the start that Eagle's Whetstones would be better than VAX's. When it became clear that the team could not match VAX's Whetstones at double precision and still keep the ALU on one board, West decided to sacrifice some performance at double precision. If they could surpass VAX at single-precision arithmetic, they could still say Eagle was faster than VAX.

In August they ran a preliminary Whetstone on an unfinished prototype. The results were not encouraging. The numbers fell below their expectations. It appeared that in single-precision arithmetic, Eagle would be no faster than VAX.

"It's just bullshit," said Jon Blau in a loud voice one evening around this time while sitting by Coke. "What's better, Crest or Colgate?"

But Rasala did not respond. "It's discouraging," he said.

"What's the problem?" I asked.

"It's somewhere back in there," he said. Without turning, he jerked his thumb over his shoulder, pointing, pointing, pointing at the prototypes that stood humming against the wall.

Allegedly, some computer companies maintain squads of debuggers called "finishers." They march into the lab and take over the job when a machine has been brought to the last stage of its incubation and the inevitable slowdown occurs. If West had owned such a team and had used them now, he would certainly have provoked a mutiny. The troops might have hung an effigy of him from the flagpole out front.

But it was time for a change, Rasala believed. He decided to work a regular shift in the lab himself. He chose Guyer to be his lab partner. It was never clear just how well Rasala did at this eleventh-hour debugging, unacquainted as he was with the intricacies of the hardware. The others had come to know it as well as you do the rooms of your childhood home; they could find all the light switches in the dark. Rasala could not, though he was spending many hours studying up on his own. Questioned about this, Guyer shrugged his shoulders and gave his high laugh. "Oh, Ed's comin' along."

Late in August, after work one day, the whole gang came trooping into the Cain Ridge Saloon, teasing Rasala by turns. It appeared that he had suspected a certain chip of overheating occasionally, and in order to find out if this was indeed the cause of the bug they were working on, he had subjected the chip to the heat gun, a glorified hair-dryer. But he had done so with too much enthusiasm. Rasala had melted the chip's socket.

The socket cost just a few pennies, but the way they carried on, you might have thought it was made of gold. The way they all laughed and cried out, Rasala cowering in mock trepidation at their abuse — "Meltdown!" "We need more fire extinguishers in the lab, Ed!" — it was clear that under his influence, they had returned.

When he first arrived at Westborough, Jon Blau heard from Rosemarie Seale that in five weeks the Eclipse Group's offices would be remodeled. It was ultimately on account of this that Blau, for one, never believed in the promise of stock options. It was like the debugging schedules. When five weeks elapsed, the promise was that the workmen had been delayed but would arrive five weeks hence — for sure this time. It became a standing joke. Of anything that clearly wasn't going to get done on time, they would say, "No sweat, we'll do it in five weeks."

Now they were back on double shifts. They had gone back underground. The Hardy Boys were attacking those bugs. The Microkids were cranking out the last parts of their code. Suddenly, the carpenters arrived. All the walls to all their offices and cubicles came down. Then the carpenters went away for a week. Rosemarie did what she could. They managed somehow. At least nothing important got lost.

"The place is all torn up," said West, "and when it's back together, I suspect that they'll find it's been emotionally torn up, too."

As it turned out, the group did not get much new space. Alsing and Rasala got walls and a real door apiece, and most of the rest got cubicles again — but ones of slightly different size, so that an engineer with some seniority, such as Holberger, got several square feet more than the recruits. Thus, for the first time, the group's pecking order was made completely visible. It was carved out in steel partitions.

The atmosphere was different. It was hard to put your finger on the change, but the veneer of equality was gone, and with it perhaps some of what had made the Eclipse Group seem unusual. Perhaps it was only my imagination, but afterward there seemed to be more talk among the younger members of the team about being "peons," more sentences in which the phrase "someone at my level" cropped up.

By early September Holberger felt he was back in "burn-out city." Around this time he found, in a wastebasket in the lab, the stub of a paycheck. It belonged to one of the technicians who had come down from Portsmouth. These were first-rate technicians, and being technicians, they got paid for overtime. Engineers were professionals; they did not.

Holberger glanced at the stub. He couldn't help it. He was astonished. The technicians were taking home more than twice as much as he was, on account of all the overtime.

Holberger took the stub to Rasala. Together, they burned it, so that the troops wouldn't see it, and then they went back to work. Holberger laughed. He said once again, "I don't work for money."

By Saturday, the fifteenth of September, Rasala felt that they were close. He inaugurated what he called "the last big push."

"We're gonna work round the clock, Saturday, Sunday," he said. "We're gonna do it all, and bathe in glory, guys." Late on Saturday afternoon, however, while running one of the last, advanced Eagle diagnostics, Gollum came down with a bad-looking bug.

The failure occurred when the diagnostic program called for an "interrupt," an order that the computer pause, store all unfinished work, and start in on another job. A computer must deal with interrupts smoothly and regularly if it is going to handle many terminals at once.

Guyer, Holberger and Rasala took on the bug. They studied the printed listing of the steps in the diagnostic program, and they saw that in addition to everything else — as if interrupts weren't tricky enough — the program was trying to test Eagle's ability to move from one block of memory to another. Then they saw that this program also contained "a page boundary crossing," another fairly arduous maneuver through memory.

Rasala felt stymied. There were too many possibilities, boundary crossings on top of block crossings on top of interrupts. "It's some complex interaction of the IP, the ATU, Microsequencer and the IOC. And probably the ALU, for that matter," thought Rasala. In effect, he was thinking, "It's the whole damn machine." So they hooked up an analyzer and they followed a signal, out of the System Cache, to Main Memory, territory into which the analyzers cannot look. And when the signal came back from Main Memory, it had changed, from a 1 to a 0 — from a high to a low voltage. Essentially, it had disappeared.

Four hours passed. Guyer left, worn out at last. Holberger and Rasala lingered awhile. They suspected it was one of those time bombs, hidden far back in the diagnostic program, but they couldn't find it and they went home. The last big push ground to a halt.

Next morning, Sunday, all three gathered in front of Gollum again. Perhaps it was a timing problem. Gollum was running with the normal, 220-nanosecond clock. Coke, next door, was still using a slow clock. They took the diagnostic program to Coke and on Coke it ran without a hitch. So it was a timing problem. Had to be. Holberger and Rasala began trading theories.

In the lab, three-way conversations often led to confusion. Guyer bowed out of this one. He pushed his chair to the table in the center of the room and rocked back. He was just thinking, gazing absently at Coke, at the boards lined up in a row inside the frame. It was like looking at a shelfful of books, and he noticed before he even realized it that Coke had one more board on its shelf than Gollum did. It had an extra memory module.

"Oh shit," said Guyer.

"What," said Rasala, without turning away from Gollum.

Guyer went leafing through the listing of the diagnostic program. "I don't wanna tell you yet," he called over. "But I think I've got it."

"You tell me now!" said Rasala.

So Guyer brought over the listing and reminded them that Coke had two boards of Main Memory, whereas Gollum only had one, and the conversation turned cryptic.

"Yup. More than two-fifty-six K," said Holberger, looking at the listing of the diagnostic.

In essence, Holberger was saying that the diagnostic programmer, in trying to test Eagle's ability to cross from one block of memory to another, had inadvertently directed the machine to cross from the block located at the very end of one board of Main Memory to the block at the beginning of the next board. But in Gollum there was no next board. So when the program told Gollum to go to the block on the board that it didn't contain, the prototype "fell off the end of the world."

The ratiocination proceeded. They had failed to recognize these facts because they were used to thinking in the numbering system that Eclipses employed. Eagle used a different system. "And we didn't change our way of thinking," said Rasala. They stuck a second board of Main Memory into Gollum and the interrupt worked. But they had lost the best part of a day.

"It's kinda embarrassing," said Guyer.

Sitting in his office, chewing up a toothpick, West scowled. Rasala should have known better.

Would West have known better?

"Maybe."

By the evening of September 20 the machines had accomplished the Eclipse Multiprogramming Reliability Test, but not quite the one designed for Eagles. Probably, Rasala thought, they would reach that milestone tomorrow. At any rate, he decided to try out Adventure the following evening.

After the M/600, his last machine, had been sent to dozens of customers, a defect had cropped up in it — an important one, which, among other things, prevented the machine from playing Adventure. If he'd tried to play the game in the lab, they'd have found the problem and fixed it cheaply. Although he had never ventured into Colossal Cave himself, Rasala resolved that he would never declare a machine debugged again until it had mastered Adventure. Now the time for Eagle to do it had come. So he thought.

When Rasala walked into the lab on the evening of the twenty-first, Holberger called out to him, "It looks like we've achieved hard interrupts." There were diagnostic programmers in the room, and some writers of system software, as well as Microkids and most of the Hardy Boys, and they were all doing something. The little room was a forest of equipment now. A couple of months had passed since Eagle had been cloned. Rasala had named the two new prototypes Tartis and Gallifrey, after the home planet and time machine of Dr. Who, the protagonist of a science fiction show on public TV. The two new machines were the first to run with the normal, full-speed 220-nanosecond clock. Like Dr. Who,

Rasala explained, the purpose of these new prototypes was "to conquer time." Mag tapes were spinning. There were disk drives everywhere. Steve Wallach was there, looking haggard, pale and excited. In addition to being emissary to Software, Wallach had become an important cheerleader. He was forever dragging important people in from the hall to show them the prototype Eagles these days — "signing up the remainder of the corporation," he said. At the moment, he was trying to get Holberger's attention, without success. Several people were trying to get Holberger's ear, and, busy as a blue jay over at Gollum, he didn't seem to be paying attention to any of them.

"Pretty exciting," I said to Wallach.

"Yeah," he said. "But there are other problems now. I'm trying to get the documentation done. Some people's writing styles are terrible. They make mistakes they should have learned to correct in *My Weekly Reader.*"

Over at Gollum, Holberger was pulling a chip out of a board with a pair of tweezers.

"Aren't you gonna take power down first?" Rasala called to him.

"No," said Holberger without looking up.

"Holberger's risking his job right here," said Rasala to the room in general, in his best deadpan manner.

There was so much equipment in the little lab now that I couldn't sort it all out. Many contraptions had their innards bared, bringing bright colors into the lab — bundles of orange and yellow wires, and brightly colored connector straps. Eleven fat logbooks for Gollum and Coke lay on the central table. Images danced on the screens of analyzers. On a table in a corner lay several printed-circuit boards. They had messages taped over their wires: DOESN'T WORK. CAN'T BE WRITTEN INTO.

A microcoder called a question to Holberger. Holberger called back, "Don't know."

"There's a rumor," declaimed Rasala, still standing in the center of the room, "that they're impeaching Holberger, to take away his Honorary Microcoder's Award."

Guyer was at the console of Gallifrey. He was trying to find the cause of a "subtle" failure in the I/O system, and to this end was trying to develop a method to increase the frequency of failures.

"That's a waste of time," said Rasala.

"No, it's not," said Guyer, and he kept on working.

Holberger was inserting his tweezers into the board of a running machine again.

"Holberger's risking his job again," called Rasala.

Suddenly, just a moment later, Holberger and several microcoders — about a third of the crowd in the lab — went charging out of the room.

"Where you goin'?" said Rasala.

"Home," said Holberger.

And a moment after that, Chuck Holland came in carrying a "disk pack"; it looked rather like a helmet. Inside it lay the code that would, presumably, enable them to play Adventure on an Eagle. Holland installed the disk pack in Gallifrey's disk drive, and Rasala sat down at the console. He typed a little, using only two fingers. His movements looked tentative. He stopped and made a blubbing sound with his tongue. Guyer pulled up his chair. Wallach, Holland and one of the diagnostic programmers approached and leaned in over Rasala's shoulder. Everyone was giving Rasala advice. It was a big moment. If it worked, if Gallifrey now opened up Colossal Cave, it would be the first time, after more than a year and a half of hard, nervous labor, that an Eagle had done something more than run tests. The computer would at last exist.

The console scratched out a message.

"Whoops," said the diagnostic programmer, looking over Rasala's shoulder. "You roached the disk."

"I told ya," said Wallach. "Ya shoulda put write inhibit."

"Now we gotta run fix-up," said Chuck Holland.

This took some time, and then the scene was repeated, all hands gathered around the console, Rasala at the controls. Rasala typed. The console scratched out a reply.

"Good-bye," said Rasala. *"Good-bye!"*

FATAL ERROR, read the message on the console.

Again, they ran "fix-up," and again the console replied, to Rasala's request to initiate Adventure, FATAL ERROR.

Rasala gave up. There would be no Adventure tonight; Eagle just wasn't ready. Rasala fed Gallifrey a diagnostic program that it had already passed, wrote up a sign, saying DO NOT DISTURB, which he placed on top of the console, and was on his way out the door, when the console started scratching out a message. So he went back. The machine had rejected its microcode. Rasala went to a filing cabinet and leaned against it. Then he took a deep breath and went back to the machine. He and Guyer made some adjustments and finally Gallifrey started running again.

"Why's it working now, all of a sudden?" I asked.

"We don't know," said Rasala. "We don't know everything about the machine."

Reaching for some comforting words, I said, "Well, at least some of it works."

"Yeah," said Guyer. "And someday we'll know which part."

Eagle was failing its Multiprogramming Reliability Test mysteriously. It was blowing away, crashing, going to never-never land, and falling off the end of the world after every four hours or so of smooth running.

"Machines somewhere in the agony of the last few bugs are very vulnerable," said Alsing. "The shouting starts about it. It'll never work, and so on. Managers and support groups start saying this. Hangers-on say, 'Gee, I thought you'd get it done a lot sooner.' That's when people start talking about redesigning the whole thing."

Alsing added, "Watch out for Tom now."

West sat in his office. "I'm thinking of throwing the kids out of the lab and going in there with Rasala and fix it. It's true. I don't understand all the details of that sucker, but I will, and I'll get it to work."

He called for Rasala one evening. "I want to go into the lab."

"Gimme a few more days," said Rasala.

On September 25 Rasala said, "As of this morning, Eagle ran Multiprogramming twelve hours overnight without failing."

Holberger said, "I know how West felt, but he couldn't have done a thing."

Back in his office West said: "It wasn't an empty threat. The game, of course, though, is that when I say I'm gonna go in there, they haul ass, because they assume it's gonna be some kind of trivial, dumb thing."

It had been such a problem, having to do with noise. But it wasn't quite over.

On October 4 most of the team clustered around Gallifrey's console, with Holberger at the controls this time, and the machine opened up Colossal Cave.

Holberger had never played Adventure before, and he wasn't about to take it on seriously now. He fiddled around in the style of a complete novice, didn't even meet the pirate or garner a treasure, and then shut the game down.

There was another game in progress that interested him more. They were running the Whetstones again.

They watched and waited most of that day. When it was all over, there was a victory. Now that Eagle was an almost fully functional machine, the figures were much better. They came out to within a few digits of what the designers had hoped for more than a year before. West's commandment was satisfied: Eagle was about ten percent faster than VAX on single precision — at least, according to the published figures for VAX — and it was about twice as fast as the fastest Eclipse.

But Gallifrey, the lead machine now, still wasn't all there. It was running all the toughest diagnostics, but failing occasionally on some of the lower-level ones. The Hardy Boys would leap for their analyzers, run the test again, and the failure wouldn't happen.

"A flake."

But where was it?

On October 6 the vice president, Carl Carman, came down to the lab as usual, and they told him about the flakey.

Carman is a man of medium height, in his forties, fair-haired, with skin rather pink from the sun — all in all cherubic-looking. He smiles like Alsing, mysteriously.

The ALU was sitting outside Gallifrey's frame, on the extender. Gallifrey was running a low-level program. Carman said, "Hmmmmm." He walked over to the computer and, to the engineers' horror, he grasped the ALU board by its edges and shook it. At that instant, Gallifrey failed.

They knew where the problem lay now. Guyer and Holberger and Rasala spent most of the next day replacing all the sockets that held the chips in the center of the ALU, and when they finished, the flakey was gone for good.

"Carman did it," said Holberger. "He got it to fail by beating it up."

They still had to prepare reams of documents. They had to run more tests. Software had to complete the vast, complex 32-bit system software. Now and then they would meet new crises, failures that usually but not always turned out to be flakey ones. They'd continue to bite their nails over the supply of PALs. Manufacturing would have to figure out how to build Eagles and the Eclipse Group would have to help. Jim Guyer still had to make the I/O system work right, and that would take a while, and in the meantime Rasala would go on saying what he had been for these many months — as if by naming the worst he could prevent it from happening: "There's still a small but finite chance that there won't be an Eagle." Nevertheless, they had reached one approximate end.

"It's a computer," Rasala said.

On Monday, October 8, a maintenance crew came into the lab with a large dolly. They loaded Gallifrey Eagle onto it carefully. They wheeled it down the hall to the Software Department. Several of the Eclipse Group walked along, as escorts, and a few of them went out that night and hoisted a few beers. But there was no ceremony this time. A few days later, in the gloom of the Cain Ridge Saloon, they shoved together some tables and held another PAL Award ceremony. Standing to give the presentation, Rasala paused, and turning to the people sitting next to him, he said behind his hand, "It's just an excuse to go out drinkin'."

Back in January Rasala had said they would open champagne when an Eagle went to Software. They had drunk champagne before, however, at less important milestones. Besides, now that the time had really come, Rasala no longer felt like drinking champagne. Partly, he was tired, more tired than he had ever felt in his life. Partly, he had begun to feel an emptiness of purpose looming; he had lived with "la machine" so long that he could not easily imagine life without it. In this mood, other problems took on perhaps an exaggerated importance. Something was going wrong, though — not with the machine, not with the group exactly — but they were in trouble.

A month or so before Gallifrey was shipped to Software, I had been chatting with Rasala in the hallway outside his office when an executive turned the corner, walking toward us. "Now we stop talking," muttered Rasala. So we did. He did not resume until the person had passed. And on the evening that I had come to see Rasala play Adventure on Gallifrey — the time the machine just wasn't up to the task — the basement had seemed steamy with intrigue. Rasala had said, cryptically, that false rumors were running around to the effect that West was on his way out.

I stopped in to see West. He had gotten his hair cut — short, like a soldier's or a respectable businessman's — and he looked plucked. He looked wan. He looked terribly thin. His face glowed slightly when I asked about the rumors. Smiling slightly, he remarked (paraphrasing Mark Twain), "The rumors of my death are greatly exaggerated."

West had said once, "Data General is a paranoid place, and the Eclipse Group does nothing to, uh, relieve that." I'd heard a story, no doubt apocryphal, about a man who while working at Data General became so fearful of his colleagues that he took to crawling under his desk in order to read his mail. I'd been having some small arguments with the company myself, and that evening I'd twisted them out of all proportion in my mind. Coming into the main lobby at Westborough, noticing the usual glances of the others sitting there and the receptionist's pleasant smile, I had the feeling that they knew something about me that I didn't know. This passed, but I felt lightened when I left the building that night with Rasala.

Rasala looked grim. I thought perhaps it was the aborted launch into Adventure, but he seemed to feel positively cheerful about that. "If it never failed, there wouldn't be any point in making it work, would there?" he said.

We went to a Burger King a few miles from the plant. While we were waiting in line, Rasala said to me, "You felt it tonight."

"What?"

"Your whole mood changed. You were relieved to get out of there."

I agreed that was so. Then I asked him if he ever felt glad to get out of there.

Rasala looked me over, up and down, as if to say it was a foolish question. "Once a night," he replied.

15

CANARDS

WEST'S RISKY DECISION to use the new chips called PALs had some troublesome consequences. Months passed before Data General could be sure of getting enough of those parts to manufacture Eagles. So the machine's public debut was put off, again and again, until the spring of 1980. As time went on, however, it became clear that West had made the right choice; PALs really were a chip of the future. Moreover, the delay gave the programmers time to create a much more impressive number of software options than usually accompanies brand-new machines. Diagnostics, meanwhile, had time to perfect a full set of microdiagnostic programs, which would help make Eagle easy to repair. There was time, too, for the Eclipse Group to refine the machine. Computers are sensitive things. With some makes, you dare not switch the boards of one with those of another supposedly identical machine. But you could switch boards among Eagles without worry by the time they were done.

A few weeks before the announcement, Eagle did come down with another evil-seeming bug. It took the Hardy Boys a long time to isolate it. They got close but could not fully identify it. They were stymied, until Holberger came up with a solution. It involved the addition of a single wire to a circuit. He said he couldn't explain why the repair would work. He just knew that it would. Rasala bet it wouldn't and as a consequence wound up buying Holberger coffee for the next couple of weeks.

Holberger made his mysterious repair over in the Software Department's area. Watching, one of the programmers declared that logic design certainly was a peculiar art. But, truly, the Hardy Boys had arrived at the state in which they could *feel* what was wrong with their machine. And there wasn't much wrong with it, by the time it finally got out the company's door.

In the months that followed Eagle's debut, it would become evident that the computer was probably going to be a big win, just as West had promised. Rumor had it that by early 1981 the dollar value of orders for Eagles represented more than ten percent of the value of all new orders for Data General equipment. So it did seem that Eagle had arrived just in time to rejuvenate the upper end of the company's product line. Data General was already late entering the supermini market, and might have missed the market altogether if it hadn't been for Eagle, because by the spring of 1981 North Carolina still had not produced a machine.

Who had been the prime mover behind this success, this act of recovery? The company's system of management? or the team itself? or West, or Edson de Castro perhaps?

Engineers in the Eclipse Group who had been around for some years still referred to de Castro as "the Captain" or as "the Man in the corner office," often in tones of amazement, admiration and fear. Eventually, some of the survivors of the team would refer to him, forthrightly, as "God."

"De Castro's down here only as a presence," West said, "yet as a presence he's an iron hand." Another time, West remarked, "De Castro does this crazy, apparently bungled stuff, but after a while you see this incredible order." And he told me: "It was de Castro who said, 'No mode bit.' It turns out that in a very succinct way he was describing the perfect machine." A succinct description indeed, if it was a description at all. West seemed to be suggesting that de Castro might have orchestrated from afar the the entire project, even including West's own conviction that Eagle's creators were working largely on their own. "I think we all have that feeling about de Castro," said a member of the team. He was only partly joking. Several engineers told me how sharp de Castro was on technical issues. Come to him with a plan that has a weak spot in it, a weak spot that you think you've cleverly disguised, they said, and he'll find it every time. You didn't put one over on de Castro. Things weren't done behind his back. Never mind how big the company had gotten, he was still in control.

Four straight-backed vinyl-upholstered chairs face de Castro's desk, which is absolutely clean, save for one small stack of papers with their edges squared. De Castro sits at a slightly lower elevation than his visitors, and when he sits down behind his desk, most of him disappears. He becomes, as it were, a sphinx — half desk, half man. He is thin and balding and wears on this occasion neither jacket nor tie. Asked about the meaning of the term *competition for resources,* he smiles broadly and says: "It's there. What it really does, in a sense, is allows for the accomplishment of certain projects that some people would prefer not to do."

Had the Eagle project always interested him or had it grown in importance gradually?

"From the start it was a very important project."

Was he pleased with the work of the Eclipse Group?

"Absolutely!" His voice falls. "They did a hell of a job."

But some members of the team felt that they had been rather neglected by the company.

"That doesn't surprise me," he says. "That's frequently the case. There's often a conflict in people's minds. How much direction do they want?"

The team seemed to think of itself as an independent, entrepreneurial outfit within the company. Did that happen by accident or had he tried to foster that feeling?

De Castro has turned his head toward a wall during a lull in the conversation. Across his face there seems to flicker a look of weariness. He turns back, smiling again. "I think that if you try to foster that attitude, you will be unsuccessful. . . . I think that from our point of view we try to provide an environment where those things oughta happen." His voice falls, and he adds, "In cases where they oughta happen."

Maybe de Castro orchestrated the project, or maybe, having put all the parts in place, he just let it happen — which might come to the same thing. Speaking privately, a veteran engineer bristled at that idea. "De Castro isn't *that* good," he said. "He's lucky and he's smart, but most importantly, he has people around him like West, who pull his ass out of a fire and then attribute it all to him." Members of the team had their own interpretations.

A month after Gallifrey Eagle was rolled slowly down the corridor to Software, Ed Rasala and several of the Hardy Boys got together at the Cain Ridge Saloon and did some reminiscing — disputatiously, as usual. At one point, Jim Guyer said: "We didn't get our commitment to this project from de Castro or Carman or West. We got it from within ourselves. Nobody told us we had to put extra effort into the project."

Ken Holberger burst out laughing.

Guyer raised his voice. "We got it from within *ourselves* to put extra effort in the project."

Laughing hard, Holberger managed to blurt out, "Their idea was piped into our minds!"

"The company didn't ask for this machine," cried Guyer. "We *gave* it to them. We created that design."

Others raised their voices. Quietly, Rasala said, "West created that design."

Rasala's big forearms rested on the table, surrounding a mug of beer. I thought perhaps I had not heard him right. He had always insisted that Eagle belonged equally to every member of the team. There was a whiff of heresy in the air.

"What did you say that West created?"

"Eagle," said Rasala.

By then the others had stopped arguing and had turned toward Rasala, who was wearing one of his looks; it seemed to warn against contradiction.

"You mean West created the excitement."

"No," said Rasala, in a flat voice. "The machine."

"The opportunity," offered Holberger.

"The machine," said Rasala.

Then there was a moment during which everyone avoided everyone else's eyes, and the conversation resumed on another subject.

Adopting a remote, managerial point of view, you could say that the Eagle project was a case where a local system of management worked as it should: competition for resources creating within a team inside a company an entrepreneurial spirit, which was channeled in the right direction by constraints sent down from the top. But it seems more accurate to say that a group of engineers got excited about building a computer. Whether it arose by corporate bungling or by design, the opportunity had to be grasped. In this sense, the initiative belonged entirely to West and the members of his team. What's more, they did the work, both with uncommon spirit and for reasons that, in a most frankly commercial setting, seemed remarkably pure.

In *The Nature of Gothic,* John Ruskin decries the tendency of the industrial age to fragment work into tasks so trivial that they are fit to be performed only by the equivalent of slave labor. Writing in the nineteenth century, Ruskin was one of the first, with Marx, to have raised this now-familiar complaint. In the Gothic cathedrals of Europe, Ruskin believed, you can see the glorious fruits of free labor, given freely. What is usually meant by the term *craftsmanship* is the production of things of high quality; Ruskin makes the crucial point that a thing may also be judged according to the conditions under which it was built.

Presumably the stonemasons who raised the cathedrals worked

only partly for their pay. They were building temples to God. It was the sort of work that gave meaning to life. That's what West and his team of engineers were looking for, I think. They themselves liked to say they didn't work on their machine for money. In the aftermath, some of them felt that they were receiving neither the loot nor the recognition they had earned, and some said they were a little bitter on that score. But when they talked about the project itself, their enthusiasm returned. It lit up their faces. Many seemed to want to say that they had participated in something quite out of the ordinary. They'd talk about the virtues of the machine — "We built it right" — and how quickly they had done it — "No one ever did it faster; at least, Data General never did" — and of the experience they had gained — "Now I can do in two hours what used to take me two days." One of the so-called kids — kids no longer, but veterans now — remarked, "This'll make my resumé look real good." But, he quickly added, that wasn't what it was all about.

Many looked around for words to describe their true reward. They used such phrases as "self-fulfillment," "a feeling of accomplishment," "self-satisfaction." Jim Guyer struggled with those terms awhile with growing impatience. Then he said: "Look, I don't have to get official recognition for anything I do. Ninety-eight percent of the thrill comes from knowing that the thing you designed works, and works almost the way you expected it would. If that happens, part of *you* is in that machine." On this project, he had reached a pinnacle the day when he finally expunged the "last known bug" from the board that he'd designed.

Engineers are supposed to stand among the privileged members of industrial enterprises, but several studies suggest that a fairly large percentage of engineers in America are not content with their jobs. Among the reasons cited are the nature of the jobs themselves and the restrictive ways in which they are managed. Among the terms used to describe their malaise are *declining technical challenge; misutilization; limited freedom of action; tight control of working patterns*. No one who made it through the Eagle project could in fairness have raised such objections. The work was divided, but it was not cut to ribbons. Everyone got responsibility for some important part of the machine, many got to choose their piece, and each portion required more than routine labor. The team's members were manipulated, to be sure, and the unspoken rules of their group were Darwinian, but many of those who made it through declared that they had been given as much freedom as they could have wished for.

Rosemarie Seale believed that West had granted them more latitude than they would have been allowed under a typical manager. She had worked for a number of other managers, and being the "mother of the team," she could speak with authority on how they had been raised:

"The bottom line on this is that effort was done; it was done well, with very little help from the corporation, if any; a lot of people were allowed to grow; a lot of people were allowed to feel good about themselves — not a pat on the back — but deep-down good about themselves. I guess all of us were trying to prove something. I was trying to prove that I could be more than a secretary, that I'm a new, liberated woman. Amazing! We all had something different to prove and we were all trying to prove the same thing.

"He set up the opportunity and he didn't stand in anyone's way. He wasn't out there patting people on the back. But I've been in the world too long and known too many bosses who won't allow you the opportunity. He never put one restriction on me. Tom allowed me to take a role where I could make things happen. What does a secretary do? She types, answers the phone, and doesn't put herself out too much. He let me go out and see what I could get done. You see, he allowed me to be more than a secretary there.

"I'm not putting Tom up as a paragon of virtue. Many's the time when I was bullshit with him. But who could expect life to be perfect? I would do it again. I would be very grateful to do it again. I think I would take a cut in pay to do it again."

A great deal has been written on the question of how to motivate industrial workers. Presumably such literature arises because so many jobs have been made so trivial that few people can find any meaning at all in them. It may be that techniques of management alone can't cure the problem. But clearly, for even the most potentially interesting jobs to be meaningful, there must be managers who are willing to throw away the management handbooks and take some risks.

Maybe in the late 1970s designing and debugging a computer was inherently more interesting than most other jobs in industry. But to at least some engineers, at the outset, Eagle appeared to be a fairly uninteresting computer to build. Yet more than two dozen people worked on it overtime, without any real hope of material rewards, for a year and a half; and afterward most of them felt glad. That happened largely because West and the other managers gave them enough freedom to invent, while at the same time guiding them toward success.

West never passed up an opportunity to add flavor to the project. He helped to transform a dispute among engineers into a virtual War of the Roses. He created, as Rasala put it, a seemingly endless series of "brushfires," and got his staff charged up about putting them out. He was always finding romance and excitement in the seemingly ordinary. He welcomed a journalist to observe his team; and how it did delight him when one of the so-called kids remarked to me, "What we're doing must be important, if there's a writer covering it."

Engineering is not of necessity a drab, drab world, but you do often sense that engineering teams aspire to a bland uniformity. West was unusual. Alsing, who might have traveled anywhere, but whose life had been largely restricted to the world of engineering, responded most strongly to him. "West," said Alsing, "took a bag on the side of the Eclipse and made it the most exciting project in the company, the most exciting thing in our lives for a year and a half. West never bored us."

As for West himself, there is no doubt that he had experienced an odd, nervous kind of fun. The circumstances had been propitious. There was a crisis. Solving it called for unorthodox procedures. He'd found, it seems to me, an opportunity to reconcile for a time two opposite ambitions — one conventional and quantifiable, the other unorthodox and vague. He was back in Cambridge, as it were, singing folk songs, while at the same time putting money on Data General's bottom line. He could be, for a little while, a balladeer of computers.

"We're building what I thought we could get away with," West had said. But Eagle sufficed. "With this machine we're going way beyond what any one person can do. I always wanted to build something like this."

Now it was done. The Eclipse Group and the many others who had worked on the machine — including, especially, Software and Diagnostics — had created 4096 lines of microcode, which fit into a volume about eight inches thick; diagnostic programs amounting to thousands of lines of code; over 200,000 lines of system software; several hundred pages of flow charts; about 240 pages of schematics; hundreds and hundreds of engineering changes from the debugging; twenty hours of videotape to describe the new machine; and now a couple of functioning computers in blue-and-white cases, plus orders for many more on the way. Already, you could see that the engineers who had participated fully would be looking back on this experience a long time hence. It would be something unforgettable in their working lives. All this, at last, was no canard.

16

DINOSAURS

MONTHS AFTER the machine's announcement, Jon Blau was cleaning up a detail; it had to do with an interconnection between Eagle's CPU and another device. Looking through some documents, he came across the name of a cable that he'd never heard of before, and puzzled, he called to Holberger, "Hey, Ken, what's this?"

"Ahhh," said Holberger, looking at the document. "That looks like a Tom West special."

At that moment Blau felt he saw it all. For the first time, he fully realized how many other groups besides his own had been brought into the project, to perform large and little tasks, such as the design and fabrication of this cable. There had been a tricky little problem here and this cable solved it. Though just a detail within a detail, it was crucial. There must have been dozens of problems like this standing between Eagle and the company's door, ones that he and the other recruits had never known existed. Who had anticipated them and arranged for their solutions? It had to have been West.

Blau knew that he could not have done that job himself. For that job, you'd need to know a lot, he thought; you'd need intimate knowledge not only of the CPU but also of the company's entire product line. He tried to imagine how hard it must have been to pull everything together. As time went on, at least one other young engineer would have an experience like Blau's; he'd come across a problem and find that West had identified and solved it long ago. As for Blau, he had often wondered what, if anything, West did behind his office door. Now he stared at the document before him, and he exclaimed, "Wow!"

There really was, as West had often said, more to building a computer than designing and debugging a Central Processing Unit. Someone had to dream up its general outlines in the first place. Someone had to make sure that the computer would work compatibly with the company's existing lines of peripheral equipment. Someone had to set goals of cost and performance and see to it that they could be met. These and many other items lay on the list of what West, often all alone, had accomplished.

Although he had rarely confessed it to anyone in the company, West had taken a large gamble. "You have to believe in yourself enough to make some pretty outrageous statements," he had said, referring to his largely successful efforts at selling the virtues of Eagle. Secretive as a mother cat about the location of her kittens, West had masked most of his activities, including his worrying.

He had once appeared to Carl Alsing as a mysterious stranger passing through town, and he really was that person — by constitution and preference a loner. He had never before owned responsibility for such a large project or for the performance of so many people. He had taken it upon himself, of course. Somehow, though, he seemed to have gotten it in his mind that he was responsible for things that he couldn't possibly control. The team had signed up to create the computer; he thought he himself had signed up to make sure that if they did their part, Eagle would get out the door and be a big success and the team would be rewarded, with stock and prestige and the freedom to play pinball again. Half the time, while sitting alone in his office, planning and planning, he had imagined that total failure was imminent. Danger made life interesting, but anxiety gets tiring after a while.

By late summer West appeared ready for the project to be over, and it nearly was for him. Late in June, having rounded up supporters in various divisions of the company, he had for the first time formally presented the machine to de Castro, at the meeting of executives called the Product Board; and while de Castro had not responded as enthusiastically as West had hoped he would, the Captain nevertheless seemed to give the machine his laconic blessing.

There had been a time when people with CB radios had been overheard in the vicinity of Westborough discussing the "battle" between the Eclipse Group and North Carolina; but nervous, invigorating talk of that sort had long since passed away. After the Product Board, Eagle was out of the closet and West slackened his public relations campaign. The machine had its own momentum now. During that summer, West suddenly remembered bike rides he'd taken with his father on Sunday evenings, and he found time to reassert that tradition with his oldest daughter. Suddenly, it seemed, he realized that his children were growing, and apparently he intended now to guide them on their way. He said one evening: "That's the bear trap, the greatest vice. Your job. You can justify just about any behavior with it. Maybe that's why you do it, so you don't have to deal with all those other problems."

I went sailing off Cape Cod with West for a few days in early August. On a morning that began with squalls and then became windless, we were heading out under the power of the boat's diesel toward the Buzzards Bay Tower, through a steep, unpleasant sea. Suddenly, the engine quit. West stared at me, with his mouth half open. Then he set his jaw and turned away.

He disappeared into the cabin and in a moment I heard some banging from below. A little later he poked his head up through the hatch and said, "Try it."

No luck. It wouldn't start. West disappeared again.

He again poked his head up through the hatch. He was holding the engine's manual aloft. "This is a good manual. The only problem is, ummmmmmmh, it's not for this engine." Then he was gone again. More banging. Try it again. No luck. West's head appeared in the hatchway, reading from the manual: " 'It is axiomatic that when a diesel engine fails while underway, that there is a fuel problem.' " He laughed, as if he had just read something funny. Again he was gone and again he reappeared, this time to spit out a mouthful of diesel fuel. Finally, of course, he made the engine run.

Back on land, one night shortly afterward, West told me that this incident was probably the closest I'd ever get to seeing him work on a computer. Then he remarked: "You assumed I could make it work. But I had never worked on a diesel engine before, and I was kinda pissed. I didn't think I could do it." I took it he was tired of playing the hero with mechanical things. Clearly, this was not the same man who would describe himself to virtual strangers by saying, "I can fix anything."

West kept on losing weight. He was tired, and he was preparing himself for a large change.

Nearly a year later, Holberger and Alsing were reminiscing about Eagle in a barroom, and Holberger remarked that what had happened near the end of the project reminded him of the typical conclusion of one sort of Western. A town hires a gunslinger to clean it up, but when he's taken care of their problem, he's still a gunslinger and sooner or later the respectable citizens are going to run him out of town.

Alsing warmed to the analogy at once. "Of course! Of course! That's a classic American story."

When the new recruits had arrived, they had been told that the Eclipse Group represented the very heart of Data General. The veterans who told them so certainly believed it. Their group, as they saw it, was the most dogged, hardworking, practical, productive and dangerous in the company, a bastion of the old successful ways, a paradigm of the company as it had been when it was small. They believed in the rule of pinball: if you win, you get to play again; but failure is unthinkable, so you'd better let no one get in your way.

Holberger's remark about gunslingers recalled some old scenes:

• I'm sitting in West's office taking notes, when I look up and see West eyeing my pad from across his desk. He smiles. Then he reads my notes to me. In business, he explains, still smiling, you learn to read memos upside down.

• West is standing at his Magic Marker board, drawing a diagram of the company's hierarchy as it affects his group. "There's some dogs and cats out here," he says. "This guy is a nonissue." He draws a large X over someone's name. "This guy disappears in time."

• Holberger sits at his terminal, sending out a bogus EMERGENCY WARNING MESSAGE.

• Rasala goes thundering into Diagnostics.

• West sits in his office and declares, "The only way I can do this machine is in this crazy environment, where I can basically do it any way that I want."

When the Eclipse Group started building Eagle, the management structure around them was extraordinarily bare. During the project, new executives were hired to fill the gaps. They came from other companies and did not know, of course, the odd, unspoken rules by which the Eclipse Group played. The team's environment was changing, and the veterans felt it, in small ways and large: Rasala complaining about a new rule that forbade employees from taking home used packing boxes without going through the formality of getting a property pass, West in a traffic jam outside the plant one morning, muttering: "This is really beginning to make me mad. I used to be able to park at the front door of this company."

One day Rasala, Holberger and West were sitting around in West's office, voicing general agreement that beating people up didn't seem to get results anymore. Most other groups just didn't seem to be willing to put in lots of overtime on a machine any-more. Picking up a line from West — that the Eclipse Group might be "a dinosaur" of a team — Holberger suggested that they order some T-shirts, which would bear the name Eclipse Group under the image of a panting *Tyrannosaurus rex*.

West and the other leaders of the Eclipse Group had acquired some enemies who would not mind seeing them taken down, and West got into a long-running battle, which heated up as the de-bugging approached completion. As he explained it to Rasala, West felt he was fighting to preserve the group's substantial free-dom and to prevent a situation in which the team would simply be delegated to do certain jobs. Rasala said: "The real fun — the way you get an Eagle out the door — is to get the guys to invent it, to come up with an innovative idea that's gonna make money. Once you've got that, there's no problem motivating people. Once we've got that, okay, baby, we're gonna do a hell of a job. And I think that's what West is fighting for."

Rasala observed the battle with mounting anxiety and then with resignation. Most people, at one time or another in their careers as adults, reenact the molting of their adolescences. Rasala liked to say that he spent his professional childhood at Raytheon, and that at Data General, under West, he grew up. For a long time he felt afraid that West would leave and that he would have to take over the group, but gradually, during many long talks with West, he became fairly comfortable with the possibility.

One day in September, at a fairly large meeting attended by bosses from outside the group, West criticized Rasala for some mistake. Feeling that he had deserved the mild rebuke, Rasala wasn't angry. Afterward, however, West took him aside. West ex-plained that he was being accused of criticizing other groups but never his own. That was why he had publicly rebuked Rasala in the meeting: he was trying to prove the allegation false. He told Rasala he was very sorry he had done it.

West, Rasala felt, was growing rather desperate, and it made him sad. At the beginning of November, Rasala remarked: "The last two weeks I've had strange conversations with West. He's beginning to teach me the loose ends. I get the strange feeling I'm in finishing school."

In retrospect, Alsing was convinced that West stayed around until the machine was safe from local politics. In November West called the team together and told them that he was leaving them; he was going to that distant country, the upstairs of Westborough, to take another job, connected with marketing, a job that would take him often to Japan. Thus he supplied them one answer to the question of what happens to computer engineers who pass forty. West said that Rasala, Wallach and Alsing would run the group now. He told them that their new projects were all lined up and assured them, "The fundamentals are right." He said he expected all to continue, as before.

Some people in the basement wore broad smiles that day and many of the team's young engineers felt indifference. To the West loyalists, it seemed as though the earth had moved.

In mid-November West went away on a business trip and Rosemarie cleaned out his office. The last item to go was the clock in the oak case. A technician who had worked under West for some years got a cart and put the clock on it. Rosemarie started leading him down the hall toward the elevator. Alsing saw them go by his office and he got up from his desk. "I'll hold the clock on the cart," he offered, "in case we go over a bump." Rasala joined the procession and so did Jon Blau.

When they got upstairs, they were lost. They had to ask directions to West's new office, from "some stranger in a suit," as Alsing put it. West's new abode was like his old one — narrow, windowless. They put the clock on the floor and then they all sat down. Rosemarie bustled out, and in a moment returned. "You guys comin'?"

Alsing said, "No, we're having a meeting in West's office."

"It's our three-o'clock Friday meeting," said Rasala.

Finally, they left. Going back downstairs, Alsing had the feeling he had just attended a funeral.

I visited West not long after that ceremony. He confessed: "I feel awful withdrawal symptoms. I wake up at three A.M., sweating, worried and so on. It's a real problem becoming just an everyday Joe." But then he got up off his couch and lingered by his fireplace, a beer in his hand. "It's really tempting for me to look back on Eagle and bask in it now," he said. He made his long "Ummmmmmmmmmh" and went on. "Next question you gotta ask yourself — is that a trap? "Ahhhhhhhhnd, the answer is: Yeah, probably."

On his coffee table, lying among heaps of magazines, I noticed, was a tall stack of books about the Orient. West was going to help show engineers at the Japanese company Data General had partly acquired how to build Data General computers. He was going to become something of a far-wandering engineer again. He was going to help launch Data General's invasion of the Far East. I thought that someone ought to warn them.

West was grinning now. He was saying: "Japan! The Orient! Man, I could really get into building machines in Japan!"

West had risen from the small death of leaving the team. In departing, Josh Rosen had felt reborn. When I looked him up many months later, I learned that he had gone to work for another computer company and was defining the architecture of another 32-bit computer. He hadn't gone to a commune, except perhaps in a comparative, figurative sense. He was working eight-hour days and five-day weeks, and he professed himself happy at last. Dave Peck quit Data General, discovering in the process that after having worked at the company practically since its founding, he was entitled to a pension: when he turned sixty-five, he would receive the grand sum of fifty-three dollars a month. Steve Wallach gave the speech he had once dreaded, describing Eagle's architecture to a jury of peers, at a meeting of a society of computer professionals, and when he was done, they got up and applauded — "the ultimate reward," he said. But Wallach left the Eclipse Group to work in another part of the company, and a year or so later he left Data General. As for Carl Alsing, he too resigned from Data General and took a job in California — "at much higher pay," he said, but he added, "That wasn't why I left." Both he and Wallach had felt unappreciated by the company. In the project's aftermath, Alsing had most of his former responsibilities taken away. "I feel," he said then, with a small smile rather like West's, "that I am free to go."

Unquestionably, Data General was changing. For one thing, the company was falling on hard times. Not long after Gallifrey Eagle was wheeled carefully down the hall to Software, Data General released a disappointing financial report, their first in many years. Profits were off. Over the next year and a half, the value of a share in Data General plunged, then rose, then fell again. More disappointing earnings reports were filed and speculation about the causes began in earnest. *Business Week* blamed de Castro's style of management mainly, but appeared not to understand that style any better than anyone else. It did seem that the company had let some of its product line become outworn, and while there was no telling just what had been lost by not fielding a 32-bit machine sooner, it was clear that without Eagle, Data General's present troubles would have looked far worse then they did.

I did not think Data General had the look of a doomed enterprise, but rather of one that was suffering harsh growing pains. During this period, what seemed to be an unusually large number of people in crucial positions left the company; they included important sales personnel and several vice presidents — among them, Carl Carman. Downstairs, remembering that the Eclipse Group had once been called "a strong foundation," one of the team asked, "If we're such a great foundation, who are we holding up?"

For at least some of the old hands, the feeling that they had gone underground to build their computer had never been stronger. Surfacing, they found the company in apparent turmoil. And what they returned to seemed nothing like the hero's welcome that they had expected.

Some of the problem was inward and inevitable. A number of the engineers confessed to feeling those long-anticipated "postpartum blues"; they spoke of feeling "that empty spot." Some of the problem lay in the local political situation. In the aftermath of West's departure, psychologists came down to visit the team and handed out questionnaires. These seemed designed, one young engineer felt, to find out what was wrong with the group, and he couldn't understand it. Why were they being asked that question, given what the team had accomplished? As for the top managers of the group, they felt that they were under attack. That situation eased within a few months, but the effects lingered. For the senior managers of the team, the project had this bitter end: that far from feeling rewarded, they thought that they were being neglected and maybe even punished for what they had done. Obviously, the company didn't intend to produce such a sour finale, but clearly those in authority didn't take in timely fashion all the right preventive steps.

Bitterness wasn't general or enduring. Many in the group got promotions, business cards, a few free trips and, though they were a very long time in coming, stock options. To their pride and delight, a number of the team's junior members got to participate in the publication of technical papers about the machine. Most important, many got to play pinball — in the sense that they went to work on those interesting projects that West had concocted and left behind for them.

In the fall of 1980 the Eclipse Group was disbanded and its members dispersed into several new and smaller groups. Some of the old crew mourned the team's passing. One said, angrily: "It was a group that was formed and achieved this remarkable thing for the company, and the company has deemed to reward that group by blowing it up. It's really sad." Many others, however, shrugged. They felt that the group's demise had been inevitable, on the one hand, and that it wasn't really dead on the other.

After the reorganization, Ed Rasala decided to leave Data General, and although the authorities tried hard to keep him, he finally did get away. He joined Alsing out west in California. West's chosen heir, Rasala was the last of the old group's managers to leave the team, and so his going had the status of a milestone.

West stayed with the company. He had never imagined that the rule of pinball would be a lie for his loyal lieutenants, and he had dreamed of the group perpetuating itself. It took him a while, but eventually he seemed to find for himself a workable attitude toward the departures of his friends and the team's demise. "Yeah, it's all blown apart, but the ethic's still in place," he said. "In some sense spreading that around may be beneficial."

West added: "It was a summer romance. But that's all right. Summer romances are some of the best things that ever happen."

EPILOGUE

LONG BEFORE IT DISBANDED formally, the Eclipse Group, in order to assist the company in applying for patents on the new machine, had gathered and had tried to figure out which engineers had contributed to Eagle's patentable features. Some who attended found those meetings painful. There was bickering. Harsh words were occasionally exchanged. Alsing, who during the project had set aside the shield of technical command, came in for some abuse — Why should his name go on any patents, what had he done? Someone even asked that question regarding West. Ironically, perhaps, those meetings illustrated that the building of Eagle really did constitute a collective effort, for now that they had finished, they themselves were having a hard time agreeing on what each individual had contributed. But, clearly, the team was losing its glue. "It has no function anymore. It's like an afterbirth," said one old hand after the last of the patent meetings.

Shortly after those meetings, Wallach, Alsing, Rasala and West received telegrams of congratulations from North Carolina's leader. That was a classy gesture, all agreed. The next day Eagle finally went out the company's door.

In New York City, in the faded elegance of the Roosevelt Hotel, under gilded chandeliers, on April 29, 1980, Data General announced Eagle to the world. On days immediately following, in other parts of the country and in Canada and Europe, the machine was presented to salesmen and customers, and some members of the Eclipse Group went off on so-called road shows. About a dozen of the team attended the big event in New York. There was a slick slide show. There were speeches. Then there was an impressive display in a dining hall — 128 terminals hooked up to a single Eagle. The machine crashed during this part of the program, but no one except the company engineers noticed, the problem was corrected so quickly and deftly. Eagle — this one consisted of the boards from Gollum — looked rather fine in skins of off-white and blue, but also unfamiliar.

A surprisingly large number of reporters attended, and the next day Eagle's debut was written up at some length in both the *Wall Street Journal* and the financial pages of the *New York Times.* But it wasn't called Eagle anymore. Marketing had rechristened it the Eclipse MV/8000. This also took some getting used to.

The people who described the machine to the press had never, of course, had anything to do with making it. Alsing — who was at the premiere and who had seen Marketing present machines before, ones he'd worked on directly — said: "After Marketing gets through, you go home and say to yourself, 'Wow! Did I do that?' " And in front of the press, people who had not even been around when Eagle was conceived were described as having had responsibility for it. All of that was to be expected — just normal flak and protocol.

As for the machine's actual inventors — the engineers — most who came seemed to have a good time, although some did seem to me a little out of place, untutored in this sort of performance. Many of them had bought new suits for the occasion. After the show, there were cocktails and then lunch, and during this time most of the engineers stuck together, like young men at a dance. At lunch, they occupied a table all their own. It was a rather formal luncheon, and there was some confusion at their table as to whether it was proper to take first the plate of salad on the right or the one on the left.

West came, too. He did not sit with his old team, but he did talk easily and pleasantly with many of them during the day. "I had a great talk with West!" remarked one of the Microkids.

He wore a brown suit, conservatively tailored. He looked as though he'd been wearing a suit all his life. He had come to this ceremony with some reluctance, and he was decidedly in the background. At the door to the show, where name tags were handed out, West had been asked what his title was. " 'Business Development,' " he'd said. At the cocktail party after the formal presentation, a reporter came up to him: "You seem to know something about this machine. What did you have to do with it?" West mumbled something, waving a hand, and changed the subject. Alsing overheard this exchange. It offended his sense of reality. He couldn't let the matter stand there. So he took the reporter aside and told him, "That guy was the leader of the whole thing."

I had the feeling that West was just going through motions and was not really present at all. When it was over and we were strolling down a busy street toward Penn Station, his mood altered. Suddenly there was no longer a feeling of forbidden subjects, as there had been around him for many months. I found myself all of a sudden saying to him: "It's just a computer. It's really a small thing in the world, you know."

West smiled, softly. "I know it." None of it, he said later, had come out the way he had imagined it would, but it was over and he was glad.

The day after the formal announcement, Data General's famous sales force had been introduced to the computer in New York and elsewhere. At the end of the presentation for the sales personnel in New York, the regional sales manager got up and gave his troops a pep talk.

"What motivates people?" he asked.

He answered his own question, saying, "Ego and the money to buy things that they and their families want."

It was a different game now. Clearly, the machine no longer belonged to its makers.

ACKNOWLEDGMENTS

MANY PEOPLE — in the end, there were hundreds — worked on the project that this book describes. I could not name them all in the text, nor does it seem just to attempt to name them in this note, for undoubtedly I would miss someone. I have not even been able to name all of the hardware engineers who participated. Several made important contributions and are not mentioned. Others who also did a great deal of the work are awarded only a small amount of space in the narrative. To all, I offer my regrets, and I hope that they do not share them. In any case, no one who reads this book should imagine that its characters make up a complete cast.

My thanks to all my friends and acquaintances in the world of computers, and to Susan T. and Muffet for help and hospitality. For their editorial assistance and encouragement, my thanks to Upton Brady, Peter Davison, Robert Manning, Sue Parilla and Michael Brandon. Thanks for general all-around help to Maureen Brown, Avril Cornel, Nina Engelhardt, Louise Desaulniers, Natalie Greenberg and Martha Spaulding; and for special assistance, to Paul Rich. Thanks for listening, to Stuart Dybek, Jon A. Jackson and Mike Rosenthal; and for their excellent, timely counsel, thanks to Messrs. Rob Duggan, George Hall, Greg Pilkington, Ike Williams and Tim Rivinus; and for painstaking work on my behalf, my thanks to Mark Kramer. And many thanks, of course, for all of the above and more, to my parents and to Fran.